Regional Solutions to Water Supply Provision

Awwa Research Foundation

About the Awwa Research Foundation

The Awwa Research Foundation (AwwaRF) is a member-supported, international, nonprofit organization that sponsors research to enable water utilities, public health agencies, and other professionals to provide safe and affordable drinking water to consumers.

The Foundation's mission is to advance the science of water to improve the quality of life. To achieve this mission, the Foundation sponsors studies on all aspects of drinking water, including supply and resources, treatment, monitoring and analysis, distribution, management, and health effects. Funding for research is provided primarily by subscription payments from approximately 1,000 utilities, consulting firms, and manufacturers in North America and abroad. Additional funding comes from collaborative partnerships with other national and international organizations, allowing for resources to be leveraged, expertise to be shared, and broad-based knowledge to be developed and disseminated. Government funding serves as a third source of research dollars.

From its headquarters in Denver, Colorado, the Foundation's staff directs and supports the efforts of more than 800 volunteers who serve on the board of trustees and various committees. These volunteers represent many facets of the water industry, and contribute their expertise to select and monitor research studies that benefit the entire drinking water community.

The results of research are disseminated through a number of channels, including reports, the Web site, conferences, and periodicals.

For subscribers, the Foundation serves as a cooperative program in which water suppliers unite to pool their resources. By applying Foundation research findings, these water suppliers can save substantial costs and stay on the leading edge of drinking water science and technology. Since its inception, AwwaRF has supplied the water community with more than $300 million in applied research.

More information about the Foundation and how to become a subscriber is available on the Web at **www.awwarf.org**.

Regional Solutions to Water Supply Provision

Prepared by:
Robert S. Raucher, **John Cromwell**, **Jim Henderson**, and **Carolyn Wagner**
Stratus Consulting Inc., P.O. Box 4059, Boulder, CO 80306-4059

Scott Rubin
Public Utility Consulting

James Goldstein and **Annette Huber-Lee**
Tellus Institute/Stockholm Environmental Institute

Charles Young
Natural Heritage Institute

and

Greg Characklis and **Brian Kirsch**
University of North Carolina

Jointly sponsored by:
Awwa Research Foundation
6666 West Quincy Avenue, Denver, CO 80235-3098

and

U.S. Environmental Protection Agency
Washington, D.C.

Published by:

Awwa Research Foundation

American Water Works Association

IWA Publishing

DISCLAIMER

This study was jointly funded by the Awwa Research Foundation (AwwaRF) and the U.S. Environmnetal Protection Agency (USEPA) under Cooperative Agreement No. R829679-01. AwwaRF and USEPA assume no responsibility for the content of the research study reported in this publication or for the opinions or statements of fact expressed in the report. The mention of trade names for commercial products does not represent or imply the approval or endorsement of AwwaRF or USEPA. This report is presented solely for informational purposes.

CONTENTS

TABLES

FIGURES

FOREWORD

The Awwa Research Foundation (AwwaRF) is a nonprofit corporation that is dedicated to the implementation of a research effort to help utilities respond to regulatory requirements and traditional high-priority concerns of the industry. The research agenda is developed through a process of consultation with subscribers and drinking water professionals. Under the umbrella of a Strategic Research Plan, the Research Advisory Council prioritizes the suggested projects based upon current and future needs, applicability, and past work; the recommendations are forwarded to the Board of Trustees for final selection. The foundation also sponsors research projects through an unsolicited proposal process; the Collaborative Research, Research Applications, and Tailored Collaboration programs; and various joint research efforts with organizations such as the U.S. Environmental Protection Agency, the U.S. Bureau of Reclamation, and the Association of California Water Agencies.

This publication is a result of one of these sponsored studies, and it is hoped that its findings will be applied in communities throughout the world. The following report serves not only as a means of communicating the results of the water industry's centralized research program but also as a tool to enlist the further support of the nonmember utilities and individuals.

Projects are managed closely from their inception to the final report by the foundation's staff and large cadre of volunteers who willingly contribute their time and expertise. The foundation serves a planning and management function and awards contracts to other institutions such as water utilities, universities, and engineering firms. The funding for this research effort comes primarily from the Subscription Program, through which water utilities subscribe to the research program and make an annual payment proportionate to the volume of water they deliver and consultants and manufacturers subscribe based on their annual billings. The program offers a cost-effective and fair method for funding research in the public interest.

A broad spectrum of water supply issues is addressed by the foundation's research agenda: resources, treatment and operations, distribution and storage, water quality and analysis, toxicology, economics, and management. The ultimate purpose of the coordinated effort is to assist water suppliers to provide the highest possible quality of water economically and reliably. The true benefits are realized when the results are implemented at the utility level. The foundation's trustees are pleased to offer this publication as a contribution toward that end.

Walter J. Bishop
Chair, Board of Trustees
Awwa Research Foundation

Robert C. Renner, P.E.
Executive Director
Awwa Research Foundation

ACKNOWLEDGMENTS

This report was funded by the Awwa Research Foundation (AwwaRF). The project was managed and administered by Mr. Ryan Ulrich and Ms. Susan Turnquist. The researchers extend their great appreciation to AwwaRF, and to Ryan Ulrich and Susan Turnquist for their management and encouragement. We also wish to thank the project advisory committee (PAC), which provided many very useful comments, guidance, support, and critiques of this project. The PAC helped shape and improve this effort. The PAC members were:

Andrew DeGraca (San Francisco Public Utilities Commission)
Thomas Grubbs (USEPA, Office of Ground water and Drinking Water)
Jon Kennedy (Tampa Bay Water)

The researchers also wish to acknowledge the useful input and support provided by several individuals who helped write and/or review several portions of this work at various stages of its development, including all the participants at the Workshop, held in Quebec City, on November 9, 2005. In addition, we would like to thank the following participating utilities and organizations, and individuals from these organizations who made key contributions: Central Puget Sound Water Suppliers Forum (Paul Fleming and others), Champlain Water District (Jim Fay and colleagues), Lehigh County Authority (Aurel Arndt and colleagues), Massachusetts Water Resources Agency (Steven Estes-Smargiassi), City of Portland Water Bureau (Mark Knudson) and Tualatin Water (Greg DiLoreto).

It also is important to credit individuals from the research team for their specific key contributions. The following members of the research team made notable contributions in the following portions of this report: John Cromwell, Scott Rubin, and Jim Henderson (the decision analysis tool, and associated chapters 5 and 6), Greg Characklis (North Carolina case study), James Goldstein (MWRA case study), Marca Hagenstad (Champlain and Puget Sound case studies), Annette Huber-Lee and Chuck Young (Portland regional case study), and Carolyn Wagner (MWD and Tampa Bay Water case studies).

Last, but not least, the researchers wish to thank the many individuals at Stratus Consulting who helped execute the technical tasks and produce this report, especially Diane Callow, Erin Miles, and Laura Cross.

ACKNOWLEDGEMENTS

EXECUTIVE SUMMARY

This report for the Awwa Research Foundation (AwwaRF) examines the ways in which "regional solutions" may provide viable and advantageous approaches to addressing several of the challenges facing the water supply community in North America and around the world. It provides a discussion of what "regional solutions" mean, identifies some of the key challenges facing water suppliers that regional approaches may help address, describes the main forms of regional agreements, and distills lessons from a diverse series of case studies of attempted and, in many instances, successful regional collaborations.

OBJECTIVES AND PERSPECTIVES OF THIS RESEARCH

A variety of water quantity and water quality challenges have increased the interest in examining regional approaches to water supply provision, and there are many different ways or options under which regional approaches might be developed and implemented. The core objectives of this research report are to help water agencies and other interested parties:

1. Recognize the full range of regional solution options
2. Understand better when, where, and why some of these regional solution options may be worth considering (given the specific challenges and context within which their utility and region operate)
3. Focus on the subset of regional solution options that may be best suited to their local circumstances and needs
4. Review each relevant option's respective benefits (pros) and issues (cons)
5. Identify implementation and design factors and approaches that can help promote success (and avert problems), if a specific regional approach is pursued

Based on these objectives, it is important to recognize that the use of terms such as "success" or "failure" does not refer to the presence or absence of a regional approach. Moving forward with a regionalized solution is not necessarily a "success" for all systems or circumstances, and this report does not advocate either for or against regionalization. Rather, success or failure applies here to whether a utility or community: (1) actively considers regional approaches, (2) makes a prudent and informed decision about whether a regional approach would be advantageous and, if so, then (3) moves forward and forges an appropriate working agreement with its neighbors—where that agreement is written in a way that helps ensure that the community realizes the benefits, and manages the issues (contains the cost) of the cooperative venture.

This research effort also attempts to embrace a range of perspectives for different readers and users of this document—it is aimed at assisting professionals from both small and large water supply systems.

MOVING FROM A TRADITIONALLY LOCAL SERVICE TO A REGIONAL SCOPE

Up to and including much of the 20th century, most water supply services were provided on a localized scale. As settlements, farm communities, and urban centers became established, water was recognized as a local need that typically was met with local resources and delivered through local institutions. However, as the U.S., Canada, and many other nations grew in population and economic activity, water supply provision became more complex. With growth came large increases in demands for water, meaning that locally available water sources in many locations were no longer adequate to meet current or anticipated needs. In many areas it became evident that water supply provision would be better served by regional approaches that better matched the geographic scope of the water resources being tapped, or that merged economic capabilities as necessary to finance the large capital outlays required to import water over long distances and develop needed storage.

Regulatory compliance issues also contribute to the growing interest in regional approaches by creating technical, management, and financial (TMF) challenges for Community Water Systems (CWSs). These challenges often are more acute in relatively small CWSs because they do not have access to the economies of scale generally applicable to the treatment technologies, or struggle to obtain the specialized operating expertise the treatment technologies require, or both. This has led many water professionals to note that regional approaches might provide opportunities through which small utilities could pool their efforts with neighboring utilities to facilitate compliance with Safe Drinking Water Act (SDWA) requirements and, in many instances, significantly reduce their costs.

DEFINING "REGIONAL SOLUTIONS"

Regional solutions, or "regionalization," is also referred to frequently as "consolidation" or "restructuring." For the purposes of this research report, these terms can be used interchangeably to refer in a general sense to any of the many options through which one water system will work with (or join) one or more water systems to address a problem or common need.

Thus, regional solutions are considered broadly here, referring to any form of cooperation across water systems to improve service and efficiencies (including reducing costs). This can embrace everything from complete physical system consolidation and interconnection at one end of the spectrum (i.e., two or more systems joining to become one single physical and management entity), to simple cooperative planning and management activities arranged by wholly independent systems (e.g., joint purchasing agreements to realize quantity discounts, or a simple "good neighbor" policy of providing mutual assistance when special needs arise) at the other end of the spectrum. This report also examines the range of physical and nonphysical approaches that reside between these extremes.

RESEARCH APPROACH AND OUTPUTS

Key parts of this research report, based on case studies and a literature review, are used to identify and evaluate the broad array of potential regional solution options. A wide variety of options are covered so that users can recognize the breadth of possible choices that are available for consideration. The intent also is to help users understand the pros (benefits) and cons (issues) of the various options, and to share key lessons learned from the case studies regarding how to enhance the likelihood of success by using specific design or program elements in implementing various approaches.

In addition, the research team has developed a computerized "decision framework tool" to help users assess their own utility's needs and priorities, within the context of their regional neighbors. The intent and design of this computerized tool (provided on an accompanying CD) is to help utilities narrow down the long list of options they examine, so they can focus on the subset of regional solution choices that appear most suitable to their circumstances.

CHALLENGES FACING WATER SUPPLIERS

There are growing challenges that utilities face in meeting their objectives, and regional approaches may provide solutions to many of these challenges. Sustainability is the core objective of water utilities, and sustainable water utilities are systems with the long-term ability to provide adequate water service while adapting to new regulations and customer demands (NRC 1997). Instead of focusing primarily on day-to-day operations, sustainable systems implement a comprehensive approach that extends to the long term.

It is increasingly difficult for water utilities to meet their goal to provide high-quality service on a sustainable basis within the landscape of their current organizational and institutional arrangements. This sets the stage for examining regional solutions, by articulating what it is that water agencies are striving to accomplish and deliver, and why changes in the scale and scope of their organizational and institutional approaches to service delivery may be an option for them to consider seriously.

The various forms of regional approaches can offer many advantages to water systems and their customers, and there are numerous examples where some form of cooperation or consolidation has provided many important benefits, and worked to the general satisfaction of all parties involved. However, regional approaches are not yet very widespread across the North American landscape, even though more than 30 years have passed since the SDWA was first enacted. This might imply one of two things:

1. Regional approaches have many real drawbacks that have kept them from more widespread acceptance and implementation
2. Misconceptions about the problems created by regional approaches (relative to any benefits to be gained) are so widespread that they have not gotten a fair hearing at the community level

The truth appears to lie somewhere in-between these two extreme perspectives. While consolidation can offer sizable benefits and advantages to many utilities, there also are numerous potential barriers, including technical, economic, institutional, and social constraints. In addition, "regionalization will be successful only when communities feel it is in their best interest and that they will have appropriate access to the utility and influence on its decision-making" (NRC 2002, p. 90).

Utilities are thus faced with both challenges and opportunities when considering regional options. Water systems and other interested parties need to evaluate objectively the full range of benefits, and also acknowledge and consider the full range of issues (potential costs), while weighing their options about potential regional approaches. This report provides a perspective on issues raised by regionalization in the form of a high-level overview of the general pros and cons (where "pros" refer to benefits that may arise in a win-win arrangement, and "cons" refer to issues, barriers, or concerns that need to be considered and that often can be addressed by developing a suitable agreement between partnering systems).

A BROAD MENU OF REGIONAL SOLUTION OPTIONS

There is a broad menu of regional solution options available for water supply consideration, and past case studies and the relevant literature offer several typologies for, examples of, and lessons learned from regional solution alternatives. There are many ways of categorizing regional water service options, and the literature often has made a distinction between physical interconnection and options that do not entail physical links between water systems. Regardless of how the options are labeled and categorized, there is a broad array of possible choices, and each tends to offer distinct advantages and disadvantages, depending on the specific circumstances a given utility faces.

The literature also reveals how regional approaches have been viewed by the public for providing local services other than water (e.g., regional solution efforts in education or police or fire protection, and associated public reactions and concerns). In fact, public attitudes about regionalization efforts in sectors other than water reveal the same challenges and perceptions that frequently are found as barriers in water regionalization efforts—typically, citizens tend to believe that they will not see an improvement in service (and many in fact worry about a loss in the quality of service), while at the same time much of the public does not expect to realize the cost savings they often are promised.

MANY ELEMENTS TO CONSIDER IN A REGIONAL AGREEMENT

The key for water agencies and other interested parties is to identify which regional options (if any) appear best suited for their specific needs, and then to forge an agreement with their neighbors so that the benefits can be realized and potential concerns addressed. This report guides water agencies and other concerned parties through this process, so that they can better understand their options, narrow down the range of options to those approaches that are likely to be best suited to their circumstances, and forge an agreement of the type that will most likely generate benefits and address their concerns.

Among the challenges to be addressed in a regional approach is the broad suite of key "elements" associated with regional solution approaches that need to be recognized and addressed. These key dimensions include: (1) determining asset ownership; (2) evaluating who should pay for various types of services, equipment, and facilities; (3) understanding and apportioning legal liabilities and regulatory accountability; (4) addressing concerns about local control and autonomy; (5) investigating operational issues associated with distribution systems, potential water quality degradation, and the chemical compatibility of treatment methods; (6) studying engineering and hydrologic issues, such as water resource constraints, water rights, and pumping; (7) addressing past and future financing, including bonds, notes, grants, developer contributions, and other sources of capital funding acquisition; (8) developing exit strategies; and (9) quantifying the level of cost savings or other benefits.

Water agencies need to recognize these and other myriad issues that typically accompany regional approaches, so that they can be better prepared to address the wide-ranging implications of regional approaches when they become engaged in evaluating or promoting regional solutions. The degree to which some of these elements may arise will depend on several circumstances, such as the type of regional approach being contemplated (e.g., informal sharing agreements raise fewer of these issues than would a formal merging or transfer of assets), as well as local conditions (e.g., whether water rights issues and/or interbasin transfers are involved).

In all instances, however, the same key theme emerges. Water systems and other stakeholders need to:

1. Be alert to identifying and considering a wide range of potential implications
2. Perform a suitable amount of due diligence so that the relevant issues are identified and considered
3. Try to address all relevant issues through a written agreement that provides suitable degrees and assignments of control and authority, so that the parties to the agreement are appropriately protected (i.e., have their concerns and issues managed) while simultaneously enabling the parties to realize the important benefits that the regional solution can offer

The key is thus to recognize and forge the right types of agreements. This applies to new regional arrangements that are being contemplated, and also to existing regional approaches in which in-place contracts and agreements may need to be modified to reflect changing circumstances, emerging issues, and increased awareness of the broader suite of implications and opportunities afforded by the regional alignment.

A USEFUL TYPOLOGY OF REGIONAL SOLUTION OPTIONS

Regional solutions can take many different forms and can be classified according to many alternative criteria. They can be classified in terms of the degree of physical interconnection or integration, or they can be classified in terms of the degree of organizational change and/or in terms of the degree of change in ownership of resources or assets involved. This broad range of possibilities and nomenclature can act as a barrier to a balanced consideration of regional solutions.

A concise "typology" of regional solutions is developed that condenses the range of categorization constructs down to six generic categories of regional solution "agreements" that can be forged between water systems. These agreements become the focal point, as they serve as the mechanism through which the key elements of a regional approach are identified and addressed. In other words, the type and content of these agreements forge the basis for helping to ensure that the benefits of a regional approach can be realized, and that the issues or concerns associated with the approach also are recognized and managed in a suitable manner.

These six types of regional agreements are distinguished from each other along a gradient extending from sharing and collaboration at one end of the spectrum, to consolidated ownership and operations at the other extreme. Our typology of regional solutions consists of the following:

- Mutual Aid Arrangements
- Sharing Arrangements
- Water Purchase Arrangements
- Collaborative Water Resource Development
- Contract Services Arrangements
- Consolidation

The essential differences among the categories are anchored in what the participants are committing to do together. As a best practice, these commitments are documented in written agreements even at the sharing end of the spectrum, where informal collaboration may already exist. A written agreement provides a number of benefits, including:

- It confirms mutual commitments and therefore provides clarity and confidence in the relationship.
- It provides a means of confirming that the "win/win" benefits envisioned by the parties are understood and that there is a commitment to deliver those benefits.
- It provides a framework within which to explicitly identify, address, and ameliorate a wide range of issues and concerns that may arise in consideration of the intended collaboration. The very process of negotiating a written agreement inaugurates a process of resolving issues and concerns.
- It provides a possible vehicle for expanding the possibilities for further collaboration down the road, by initiating a practice of negotiating collaborative agreements between partnering utilities. Ultimately, recognizing that an agreement is a control instrument can go a long way toward alleviating concerns regarding who is in control of collaborative arrangements.
- It offers a process (i.e., negotiating an agreement) that should expose the problems, pitfalls, and concerns that may arise in a regional collaboration. If these issues can be resolved through negotiations and by making provisions to address them in the agreement, then they are much less likely to present problems later on and can enable all participants to proceed with greater confidence. If issues and concerns cannot be resolved through the process of negotiating an agreement, it is a sure sign that the deal being discussed is not a viable collaboration.

A key point is that negotiations leading up to a potential regional collaboration agreement provide a means of testing the proposed collaboration. There is benefit even when negotiations fail since it confirms that a beneficial win/win opportunity has not been missed, but was not really as beneficial as it seemed and was not the right choice. Negotiations also advance a dialogue between neighbors that may produce alternative ideas for collaboration or improve the possibilities for collaboration down the road as circumstances change.

Finally, it should be recognized that many utilities may engage in multiple types of agreements. For example, a utility may have water purchase agreements established with some neighboring systems, and mutual aid agreements on other issues in effect with the same or other utilities in the region.

A DECISION FRAMEWORK TOOL

As part of this research effort, a Decision Framework Tool (a computerized spreadsheet model) has been developed and is provided on an accompanying CD. The decision framework contains three primary steps. Step 1 is a self assessment of the user's utility, and the region, to identify needs, challenges, and opportunities for the system (this also can be used to evaluate neighboring systems). Step 2, an automated evaluation, identifies and ranks (based on the responses in Step 1) priorities for the user's system (or the neighboring system). In Step 3, the tool directs the user to a focused subset of regional agreement options that best match the system's needs and priorities.

CASE STUDIES AND KEY LESSONS LEARNED

Several case studies have been developed by the research team, in concert with the participating utilities. These provide examples of the motivations for and types of regional approaches that have been attempted and, in most cases, successfully implemented within the U.S. The key points for each case study include the motivation for exploring regional approaches, what regional approaches were explored and adopted (or rejected), key factors contributing to the implementation (or lack thereof) of the regional approach, key barriers to be overcome (and how they were or were not surmounted), and major lessons learned.

Overall, case studies provide a method to distill central themes, identify user tips, and describe key lessons learned from actual utility experiences. Key observations and conclusions drawn from the case studies include: (1) successful regional approaches are often driven by a clear recognition that a significant problem exists and that regional approaches can help solve it (and, the problem that predominantly leads to regional collaboration is a clear lack of adequate water supply), (2) regional approaches appear more palatable if each party retains ownership of certain key assets (e.g., local distribution systems and retail customers), (3) it is useful to have well articulated benefits (e.g., estimated cost savings) that will be derived from a regional approach, (4) regional entities need to have sufficient authority to govern themselves and gain access to sufficient capital to finance investments in water and infrastructure that are necessary for sustainability, (5) it is often vital to have a champion to initiate and follow through on regional collaboration efforts—a person who, through personal drive, perseverance, and knowledge can forge and sustain the trust and relationships needed to move ahead, (6) the availability of technical and/or financial assistance (or regulatory mandates) from higher levels of government (e.g., state or federal agencies) is a very helpful promoter of regional approaches (and, likewise, having strong directives such as Court orders or groundwater withdrawal limits from higher levels of government can spur action), and (7) retaining some measure of local autonomy is a critical issue that frequently determines whether a regional approach will be considered favorably (and, regional approaches can be structured in ways that help ensure—rather than threaten—local autonomy).

CHAPTER 1
INTRODUCTION

This report for the Awwa Research Foundation (AwwaRF) examines the ways in which "regional solutions" may provide viable and advantageous approaches to addressing several of the challenges facing the water supply community in North America and around the world. This introductory chapter provides a brief discussion of what "regional solutions" means, identifies some of the key challenges facing water suppliers that regional approaches may help address, describes the objectives of this research report, and provides an outline of the content and structure of the remainder of this report.

BACKGROUND

The provision of water supply services traditionally has been developed on a fairly localized scale—especially through most of North America. In most instances, public water supplies have been drawn from local surface or groundwater sources, and treated and distributed by entities (i.e., water supply utilities) that are also very localized (e.g., owned and operated as a part of the local municipal government, or by a local developer or homeowners association). This very localized approach made sense as North America's population and level of economic activity grew over the past century. As settlements, farm communities, and urban centers became established, water was recognized as a local need that typically was met with local resources and institutions. As a consequence, there are approximately 53,000 Community Water Systems (CWSs) in the United States (U.S.) today (a CWS is defined as a system providing water to 25 or more people, or 15 or more service connections, on a regular basis). Of this total, about 84% (nearly 45,000) are classified as "small" systems serving fewer than 3,300 persons (~1,000 service connections). Two-thirds of these, about 30,000, are classified as "very small" systems serving fewer than 500 persons (USEPA 2002a).

As the United States, Canada, and many other nations progressed through the twentieth century, however, the issue of water supply provision became more complex. With population and economic growth came large increases in the demands for water, meaning that locally available water sources in many locations were no longer adequate to meet current or anticipated needs. Because water resources are often regional in nature—due to the broad geographic scope of river basins and aquifer systems, and the hydrologic interconnections between them—increased extraction in one location has often revealed adverse implications in another. Thus, it has become evident in many areas that water supply provision might be better served by considering regional approaches that more suitably match water extractions with the geographic scope of the water resources being tapped. Also, as some areas looked to tap distant water resources to meet growing demands, multiple local entities within the water importing region have needed to join forces in order to finance the large capital outlays associated with long distance water conveyance and storage.

Other factors also have contributed to growing interest in regional approaches to water supply provision. For example, increased water quality regulation under the Safe Drinking Water Act (SDWA)—signed into law in 1974, and amended in 1986 and 1996—creates technical, management, and financial (TMF) challenges to water suppliers. These regulatory compliance challenges are especially acute for the many small CWSs scattered through North America. This is because meeting new drinking water standards, such as Maximum Contaminant Limits (MCLs) issued under the SDWA by the U.S. Environmental Protection Agency (USEPA), requires treatment technologies that often are expensive and complicated. These challenges thus are more acute in relatively small CWSs because they do not have access to the economies of scale generally applicable to the treatment technologies, or struggle to obtain the specialized operating expertise the treatment technologies require, or both. This has led many water professionals to note that regional approaches might provide opportunities through which small utilities might pool their efforts with neighboring utilities in order to facilitate compliance and, in many instances, significantly reduce the cost of doing so.

These water quantity and water quality challenges, and other challenges as well, have increased the interest in examining regional approaches to water supply provision. There are many factors in addition to those noted above that are sparking this interest, and there are many different ways or options under which regional approaches might be developed and implemented. The balance of this report focuses on (1) describing the various options available for regional solutions, and (2) providing information and tools so that water agencies and other interested parties can better understand the pros and cons of the various alternatives available for their consideration, and chart a pathway to regional agreements if they are deemed suitable.

DEFINING "REGIONAL SOLUTIONS"

Regional solutions, or "regionalization," is also referred to frequently as "consolidation" or "restructuring." For the purposes of this research report, these various terms can be used interchangeably to refer in a broad and general sense to any of the many options through which one water system will work with (or join) one or more other water systems to address a problem or common need. Grigg (1989) likewise defines regionalization of water supplies broadly as "integration or cooperation on a regional basis."

There are numerous similar definitions for what is meant by consolidation or regionalization. According to the National Research Council (NRC), water system consolidation can be defined as one CWS being absorbed into, combined with, or served by other utilities to gain the resources they lack otherwise (NRC 1997). Subsequently, NRC further observed that regionalization "includes the combination of utility organizations, wholesale service arrangements, cooperative agreements, and satellite management of multiple systems" (NRC 2002, p. 4), as well as public or private partnerships, water supply agreements, system interconnection, and water wheeling.

As noted within the generalized definitions of the term, there are many possible forms of regional consolidation. These many forms of regional approaches have often been discussed in two very broad categories relating to the presence or absence of physical integration:

- The first category includes those regional approaches that entail actual *physical interconnection or integration*, such as where two or more CWSs share a common source water or treatment plant (or both) or have their distribution systems interconnected.*
- The second category includes those regional consolidation approaches that involve only *nonphysical arrangements*, such as sharing arrangements or satellite management

---

* In some literature, regional approaches using physical interconnection are classified as "structural" approaches. However, for our purposes, the "structural" label is not used, as it may confuse readers because there are many options that involve changing the structure of organizational relationships, even where no physical interconnection is made.

approaches, wherein the two (or more) water agencies remain distinctly separate in a physical sense.

Under either type of arrangement, utilities can enter into mergers or other cooperative agreements with other systems, or transfer management and/or ownership to another entity (NRC 1997). In other words, the distinction between the presence or absence of physical integration does not limit the discussion about how regional approaches may entail changes in ownership or changes in organizational structure. The options for changing organizational arrangements constitute an additional dimension of how "regional solutions" are defined.

For the purposes of this report, regional solutions are considered broadly, referring to any form of cooperation across water systems to improve service and efficiencies (including reducing costs). This can embrace everything from complete physical system consolidation and interconnection at one end of the spectrum (i.e., two or more systems joining to become one single physical and management entity), to simple cooperative planning and management activities arranged by wholly independent systems (e.g., joint purchasing agreements to realize quantity discounts, or a simple "good neighbor" policy of providing mutual assistance when special needs arise) at the other end of the spectrum. This research report also examines the range of physical and nonphysical approaches that reside between these extremes.

While our approach looks at consolidation in a broadly defined manner, in some instances the research emphasizes options that entail some form of physical interconnection or integration for purposes of water resource sharing. This is because in many of the most challenging circumstances that might spur consideration of regional solutions, such as water resource supply limitations or regulatory compliance issues, it is the physical sharing of capital and/or water resources that would most likely solve the problem or significantly reduce the cost of doing so. However, it is important to recognize that there also can be considerable efficiencies gained and costs saved through nonphysical forms of consolidation, such as cooperative ventures for joint purchasing, shared administration, or satellite management.

In sum, regional solutions can take many different forms and can be classified according to many alternative criteria. They can be classified in terms of the degree of physical interconnection or integration. They can be classified in terms of the degree of organizational change and/or in terms of the degree of change in ownership of resources or assets involved. This broad range of possibilities and nomenclature can act as a barrier to balanced consideration of regional solutions. As a product of this research, we present a "typology" of regional solutions in chapter 5 that attempts to streamline the process by reducing the range of categorization constructs down to six generic categories of regional solution agreements.

OBJECTIVES AND PERSPECTIVES OF THIS RESEARCH

The main motivations and objectives of this research project are to help water agencies and other interested parties:

1. Recognize the full range of regional solution options
2. Understand better when, where, and why some of these regional solution options may be worth considering (given the specific challenges and context within which their utility and region operate)

3. Focus on the subset of regional solution options that may be best suited to their local circumstances and needs
4. Review each relevant option's respective pros and cons
5. Identify implementation and design factors and approaches that can help promote success (and avert problems), if a specific regional approach is pursued

In compiling the information presented in this report, the research team has attempted to embrace a range of perspectives for the various possible readers and users of this document. Toward that end, the information is intended to assist professionals from both small and larger water supply systems. Regionalization often has been seen as predominantly a small systems issue, but in many instances larger-scale metro-area regional approaches (e.g., with multiple municipal and other independent local government entities within a large metropolitan area) have been valuable. Thus, we attempt to offer information that applies to both small and large systems, and that spans regions of potentially different population or geographic size.

Also, this report aims to address a range of applicable perspectives on the "roles" that different utilities might play in regionalization efforts and deliberations. For example, large utilities (or even some smaller systems) may, in effect, be "sellers" of regionalization approaches with their neighbors (i.e., some systems may be looking to sell regional approaches to their neighbors, perhaps seeking to "acquire" or simply engage neighboring utilities to work collectively). Alternatively, other water systems may be in a role as a "buyer" of a regionalization approach (e.g., a small system being approached by another, often larger system that wants the neighbor to "buy in" and join a regional approach). In a similar vein, one system may be a wholesale provider (seller) of water to a neighboring water utility in the region (buyer), and our intent is to provide information that both perspectives can use to understand their options and implement suitable regional approaches in a manner that is most likely to succeed.

Finally, we need to point out that our use of terms such as "success" or "failure" does not refer to the presence or absence of a regional approach. Moving forward with a regionalized solution is not necessarily a "success" for all systems or circumstances, and this report is not advocating either for or against regionalization. Rather, the notion of success or failure applies here to whether a utility or community: (1) actively considers regional approaches, (2) makes a prudent and informed decision about whether a regional approach would be advantageous and, if so, then (3) moves forward to realize a suitable regional approach by forging an appropriate working agreement with its neighbors—where that agreement is written in a way that helps ensure that the community realizes the benefits, and manages the issues (contains the cost) of the cooperative venture. Thus, success here is based on a system making and implementing a good choice for itself and its customers. If a regional approach is not deemed worth pursuing, then a community will succeed if it recognizes this and acts accordingly, and vice-versa.

RESEARCH APPROACH AND OUTPUTS

Our research approach consists of various elements and has generated multiple products for the end user. For example, a key part of this research report is based on case studies and a literature review. These are used to identify and evaluate the broad array of potential regional solution options. The intent is to cover a wide variety of options, so that users can recognize the breadth of possible choices that are available for consideration. The intent also is to help users understand the pros and cons of the various options, and to share key lessons learned from the case studies regarding how to enhance the likelihood of success by using specific design or program elements in the implementation the various approaches (e.g., the potential value of allowing individual entities within a regional approach to retain ownership of key assets).

In addition, the research team has developed a computerized "decision framework tool" that is intended to help users assess their own utility's needs and priorities, within the context of their regional neighbors. The intent and design of this computerized tool (provided on an accompanying CD) is to help utilities narrow down the long list of options they examine, so they can focus on the subset of regional solution choices that appear most suitable to their circumstances. The goal is to help guide users—through a computerized spreadsheet tool made user friendly with Visual Basic programming—do self assessments, identify priorities, and evaluate the relevant subset of regional solution options.

OUTLINE OF THE REPORT

This report consists of a series of chapters following this introduction. The outline of the report, and generalized content of the chapters that follow, are summarized below.

Chapter 2 provides an overview of the challenges that utilities face in meeting their objectives, and provides an introduction as to why regional approaches may provide solutions to some of these challenges. The chapter opens with a discussion of sustainability as the core objective of water utilities. The chapter then outlines many of the increasingly difficult challenges facing water utilities in their quest to provide high-quality service on a sustainable basis within the landscape of their current organizational and institutional arrangements. This sets the stage for examining regional solutions, by articulating what it is that water agencies are striving to accomplish and deliver, and why changes in the scale and scope of their organizational and institutional approach to service delivery may be an option for them to consider seriously. A perspective on issues raised by regionalization also is provided, in the form of a high-level overview of the general pros and cons (where "pros" refer to benefits that may arise in a win-win arrangement, and "cons" refer to issues, barriers, or concerns that need to be considered and that often can be addressed by developing a suitable agreement between partnering systems).

Chapter 3 provides a review of the broad menu of regional solution options available for water supply consideration. This chapter draws on the relevant literature regarding regional solutions, as well as some abbreviated case study perspectives from the literature on the past use of regional approaches in pursuit of water supply sustainability and regulatory compliance. Finally, a review is offered of lessons learned from regionalization efforts in sectors other than water (i.e., regional solution efforts in education or police and fire protection, and associated public reactions and concerns).

Chapter 4 then discusses the broad suite of key "elements" associated with regional solution approaches. This provides an overview of the many aspects of regional solutions that need to be recognized and addressed, looking beyond the engineering and hydraulics. These key dimensions include considerations such as: (1) determining asset ownership; (2) evaluating who should pay for various types of services, equipment, and facilities; (3) understanding and apportioning legal liabilities and regulatory accountability; (4) addressing concerns with local control and autonomy; (5) studying technical issues associated with distribution system operations, such as potential water quality degradation and the chemical compatibility of treatment methods; (6) studying engineering and hydrologic issues, such as water resource constraints, water rights, and pumping; (7) addressing past and future financing, including bonds, notes, grants, developer contributions, and other sources of financing; (8) developing exit strategies; and (9) quantifying the level of cost savings; and several others. The intent is to provide readers with a broad perspective of the myriad issues to consider in regional approaches, so that they can be better prepared to recognize and address the wide-ranging implications of regional approaches when they become engaged in evaluating or promoting regional solutions.

Chapter 5 provides a useful typology of regional solution agreement options. In this chapter, we distill the many different types of regional options (as initially described in chapter 3) down to a more streamlined construct that focuses on a range of six general types of "agreements" that can be forged between water systems. These types of agreement become the focal point for much of the rest of the report, as the agreements developed and entered into by water systems serve as the mechanism through which the key elements of a regional approach are identified and addressed. In other words, the type and content of these agreements forge the basis for helping to ensure that the benefits of a regional approach can be realized, and that the issues or concerns associated with the approach also are recognized and managed in a suitable manner.

In chapter 6, the decision framework tool (our computerized spreadsheet model) is described. The objectives and approach are described briefly. The process contains three primary steps. Step 1 is a self assessment of the user's utility, and the region, to identify needs, challenges, and opportunities for the system (this also can be used to evaluate neighboring systems). This leads to Step 2, an automated evaluation that identifies and ranks (based on the responses in Step 1) priorities for the user's system (or the neighboring system). In Step 3, the tool directs the user to a focused subset of regional agreement options that best match the system's needs and priorities. The chapter also provides a description of where and how to access and apply the tool (e.g., how to use the accompanying CD).

Chapter 7 provides summaries of several case studies the research team developed. These provide examples of the motivations for and types of regional approaches that have been attempted and, in most cases, successfully implemented within the U.S. This chapter provides abbreviated versions of the case studies, and the full-length case studies are provided as appendices to this report. In this summary, the research team focuses on the key points for each case study, including the motivation for exploring regional approaches in that case, what regional approaches were explored and adopted (or rejected), key factors contributing to the implementation (or lack thereof) of the regional approach, key barriers to be overcome (and how they were or were not surmounted), and major lessons learned. Overall, the intent is to distill central themes, user tips, and key lessons learned from the case studies. For example, we note that realized regional approaches have most often been driven by clear lack of adequate quantities of water supply, that regional approaches appear more palatable if each party retains ownership of certain key assets, and that retaining some measure of local autonomy is a critical issue that frequently determines whether a regional approach will be considered favorably.

Finally, in chapter 8, a summary of key guidance points, and a suggested agenda for future research is provided. This chapter opens with a brief summary of key themes and findings (an abbreviated recap of the Executive Summary). Then, descriptions are offered of future research that would be most fruitful to the community of water suppliers, and those who work with them, as they face the challenges ahead and consider the role of regional approaches as potential solutions to those challenges.

Appendices provide more complete versions of the two quantitative case studies, and a complete set of citations is provided in the References section.

CHAPTER 2
BACKGROUND ON REGIONAL SOLUTIONS

The fundamental objective of a water system is to provide safe and reliable water service at a reasonable cost. A number of operational challenges must be met day in and day out to meet this objective. More than that, however, the nature of the day-to-day challenges has changed significantly over the past few decades and will continue to change in the future, presenting a moving target. Water systems must not only meet today's needs but also be capable of adapting to meet tomorrow's needs as well.

Since the challenges that exist today are different than those that existed when the system was originally organized, the question is often raised as to whether the current form of organization is still the best choice. Regional solutions are commonly nominated for consideration in this context because so many water systems were created initially at small scale with limited capacity to adapt to change. Operating in a regional mode is viewed as a logical alternative because there are many more possibilities and resources for adapting to changing circumstances for systems at larger scale.

To properly evaluate regional solutions and related consolidation options, we need to begin with an overview of what water supply utilities strive to accomplish, what challenges they face in meeting their objectives, and what options may be available to help solve these problems. This provides an essential background on the context and motives for considering whether or not regional consolidation may make sense for a water system and its service communities.

EXISTING WATER SYSTEM ORGANIZATIONS AND CORE OBJECTIVES

Of the almost 53,000 community water supply systems regulated by the USEPA, 84% are classified as "small" systems serving fewer than 3,300 persons (~1,000 service connections). For the most part, the 16% of the industry serving more than 3,300 persons are presently benefiting from some of the advantages of larger scale operation. However, it does not preclude them from realizing additional advantages from further pursuit of regional solutions to the provision of water supply in partnership with their neighbors.

Throughout the history of many large urban water systems, a core objective has been a quest to "rationalize" the original center city system for providing water supply being coupled with components annexed during growth periods, and with historic factors affecting resource availability and means of transmission and storage. In many places, elaborate water service patterns have resulted that are inefficient to operate, unsustainable over the long term (due to unrestrained competition for common resources), and vulnerable to disruption during drought periods. Competition between urban and suburban jurisdictions has produced "water wars" in many metropolitan regions. These historic organizational schemes and water war experiences have left lots of room for potential improvement. In other regions where collaborative arrangements have instead evolved to provide wholesale development and treatment of water resources at the regional level, there is a constant challenge to hold such partnerships together as circumstances change.

Of the 44,500 "small systems" serving fewer than 3,300 persons, about 30,000 of them are classified as "very small" systems serving fewer than 500 persons (Shanahan 2002). There are tremendous differences in scale of operation and adaptability to change between "small systems" and "very small" systems. There are also many important organizational differences between systems serving more than 500 persons and those serving fewer than 500 persons. Of the 30,000 systems serving fewer than 500 people, almost 22,000 (72%) are in private ownership. In contrast, within the 500–3,300 persons served category, 70% of systems are in public ownership.

Geography is also an important characteristic of existing organizations that has a determining influence on the potential for regional solutions. There are about one and a half times as many of the "very small" water systems in suburban areas as in rural areas. Systems located in metropolitan areas are significantly closer to their neighbors than systems located in rural areas (Ottem, Jones, and Raucher 2003).

Since coping with change is such an essential organizational skill, it is instructive to review how the different ownership types that constitute the small system status quo came into being originally, their core objectives at the time of formation, and how their history may leave them predisposed to either good or bad fortunes in the future.

Government-Owned Utilities

As providers of a set of essential goods and services for the communities they serve, most local water providers have a common set of core objectives. Water suppliers strive to provide their customers with reliable delivery of safe, high quality water, in sufficient quantity to meet anticipated demands. Concurrently, most utilities try to keep their services as inexpensive and affordable to their customers as possible, while also maintaining the longer-term sustainability of the system.

Municipal water providers also may be seen as one of the core institutions available for meeting the broader goals and objectives of their service communities, playing an integral role in a process often referred to as "local control" or "local autonomy." For example, water systems may be viewed as a mechanism to facilitate or manage the form, scale, and location of future growth in the community (i.e., decisions about water and wastewater infrastructure expansion and service provision often become de facto methods of local land use planning). Water systems also may be seen as one of the tools available to the community to assure its long-term independence and self-identity. Maintaining a water system as a sustainable local entity may be seen as part of a broader strategy through which some communities seek to better assure self determination—a way to provide themselves some insulation from external trends or practices as might emerge in neighboring regions.

Investor-Owned Utilities

Investor-owned utilities share many of these objectives, but also need to be profitable for their stockholders. Private utilities, therefore, tend to emphasize growth, through a combination of new construction and the acquisition of existing water systems, more than their public owned counterparts. Understandably, investor-owned utilities also are not likely to try to use the water system to further land-use planning or other objectives that are not directly related to the operation of the water system.

"Mom and Pop" Water Companies

These are small investor-owned utilities that began independently or as part of a "company town" type of business that began and continued to grow with the town with the simple objective of filling the need for a water supply. Because the owners have limited financial resources to draw from, they can face difficult challenges in both boom and bust periods. There are many financially distressed companies in this category.

Homeowners' Associations, Mobile Home Parks, and Developer-Owned Systems

Many small, privately owned utilities were built to serve one or a few housing developments. These types of systems include homeowners' associations, mobile home parks, and developer-owned water systems. The initial, core objective of these types of systems is to provide a water supply that is sufficient to enable the sale of homes. Frequently, these systems lack long-term plans and the professional expertise to ensure the sustainability of the system's operations. After their initial objective is met (most of the homes are sold), the systems will continue to provide water service, but often without a sufficient revenue base or other resources to do so in a safe, reliable, and sustainable manner.

Rural Water Cooperatives

Rural water cooperatives have been formed as private, nonprofit entities in many places across the country to build, operate, and maintain water systems in accordance with management principles instilled as part of the U.S. Department of Agriculture's (USDA's) grant and loan programs that are today implemented by the Rural Utility Service (RUS). The financial discipline imposed by the federal programs has produced a form of co-op that has significantly better capability than homeowners associations.

Throughout the following discussion, the term "community" will be used to refer to the area served by the utility. Even where the utility is not owned by the local government, in most jurisdictions local officials have some role in determining, overseeing, or influencing the utility's actions.

CHALLENGES FACED BY COMMUNITY WATER SYSTEMS

> "The challenge facing the water industry at the onset of the 21st Century is to learn new ways to accommodate the many demands of the public, while continuing to meet society's need for adequate supplies of high-quality water."
>
> From *Managing Constraints to Water Source Development*, AwwaRF project #343 (Graham et al. 1999, p. 11).

Water utilities face a range of increased challenges in meeting their basic objectives, and meeting these objectives in a sustainable manner is especially challenging. These challenges include: (1) customer demands for more water, higher quality water, and more reliable service; (2) environmental restrictions on current and potential water sources; (3) increasingly stringent regulatory requirements (and the associated cost and technical complexity of compliance); and (4) institutional and economic constraints on operations.

Table 2.1 provides a listing of these and other common challenges facing water supply agencies. These challenges often are especially onerous for small water systems, because their limited customer base often means that they may lack the financial and other resources needed to address the increasing costs and technical complexities of sustaining operations in a changing environment. However, even larger systems face appreciable challenges that may make regional

Table 2.1
Some of the challenges for water utilities

• Ensure sufficient supply to meet growing demand for water
• Meet higher level water quality standards
• Manage external risks (weather events, security threats, chemical spills, etc.)
• Protect in-stream flows for the environment and recreational uses
• Protect endangered species
• Act as environmental steward on water issues
• Abide by local, state, and federal water rights laws
• Abide by tribal water rights and treaties
• Replace aging infrastructure
• Install new technologies
• Attract, train, and retain qualified personnel
• Implement conservation programs
• Balance competing needs and wants of agricultural, industrial, commercial, and municipal customers
• Improve reliability for customers
• Avoid raising rates or increasing taxes
• Provide enhanced security
• Upgrade emergency response plans and capabilities

solutions valuable option for them to consider. This report examines the pros and cons of various types of regional solutions that might be considered for helping to meet these challenges.

SUSTAINABILITY AS A FUNDAMENTAL OBJECTIVE AND CHALLENGE

Of all the core objectives, sustainability probably is the most fundamental consideration for local community water systems, and perhaps their largest overall challenge. Sustainability refers to developing and maintaining the long-term capability for a CWS to continue operating in a safe, reliable, compliant, and fiscally sound manner. Sustainability entails having the right types and levels of TMF resources and capabilities available to the utility.

TMF resources can be especially difficult to obtain for small rural CWSs. This is because the economic base of the community is limited by the small service populations that must shoulder the expense of costly new capital outlays and operating requirements. This problem often is compounded by the fact that many rural communities are populated by a disproportionate number of households with incomes below the national average (Rubin 2001; Ottem, Jones, and Raucher 2003). It is thus difficult to gain access to the capital required for new source development or regulatory compliance, and still operate at an annual household water bill level that is affordable for the customer base served. Likewise, it is difficult to hire and retain staff with sufficient technical and management skills for navigating the CWS through increasingly complex operating environments.

Larger water systems also face challenges to their long-term sustainability and their overall ability to serve their community and customers at the level of reliability, quality, and cost to which they have become accustomed. Challenges faced by larger communities include the following:

1. Aging infrastructure and the associated challenges of maintaining service reliability and incurring the capital expenses and service disruption costs associated with large-scale infrastructure renewal efforts
2. Increasing water demands associated with population or economic growth
3. Declining quality or availability of source water supplies (or, both demand growth and supply constraints arising concurrently)
4. Growing needs for enhanced investments in water system security and emergency power back-up
5. Increasing regulatory compliance pressures and compliance costs, to meet new federal or state (or provincial) standards (e.g., MCLs) or guidelines
6. Rising service expectations by customers for more reliable and higher quality tap water
7. Mounting political pressures to keep water rates (prices) from escalating, despite the need for substantial maintenance and compliance investments
8. Facing region-wide pressures as cities and surrounding communities grow more interdependent—in terms of regional transportation and other basic service needs—over time

These challenges are significant, and the pressures created are forcing most water suppliers—large and small—to look for ways to continuously improve their operations and capabilities. Utilities are typically looking to maintain and enhance service, and to do so in a more efficient and responsive manner.

Regional approaches can, in many instances, offer some assistance in trying to meet these objectives and challenges. However, regional solutions also impose costs and complications that should not be overlooked. The key is to understand the full range of options available, and to then recognize and weigh the full range of potential costs and benefits. This report is aimed at helping water utilities, and other interested parties, do that.

ELEMENTS OF SUSTAINABILITY

Sustainable water systems are systems with the long-term ability to provide adequate water service while adapting to new regulations and customer demands (NRC 1997). Instead of focusing primarily on day-to-day operations, sustainable systems implement a comprehensive approach that extends to the long term. Sustainable systems have the following characteristics (Wade Miller Associates 1991, Okun 1995):

- A commitment to meet service expectations
- Access to water supplies of sufficient quality and quantity to satisfy future demand
- A distribution and treatment system that meets customer expectations and regulatory requirements
- The technical, institutional, and financial capacity to satisfy public health and safety requirements on a long-term basis

A continual assessment of current and future plans, processes, skills, and services is required to identify options and implement solutions that promote sustainability (NRC 1997).

Sustainability entails looking for deficiencies within each and continually evaluating the following factors (NRC 1997):

- Capacity
- Capability
- Socioeconomic environment
- Resource environment
- Regulatory requirements
- Technical assistance available
- Financial assistance available

The NRC recommends that systems identify challenges to sustainability and respond to them using their in-house capabilities, or developing the necessary resources to do so. "Maintaining this long-term focus in the face of pressing immediate needs is one of the greatest challenges small water systems face" (NRC 1997).

Provision of capital facilities is a significant requirement in the water supply business. Water supply is inherently more capital-intensive than other utility services. Industry-wide, water utilities require twice as much capital per dollar of revenue than electric power utilities (Beecher, Mann, and Stanford 1993). In the small water system segment of the water industry, the intensity of the use of capital in relation to the scale of operation becomes even more extreme (Beecher, Mann, and Stanford 1993).

- Small water systems in the 500–3,300 persons served size category require about four times as much capital per gallon of water sold as systems serving more than 50,000 persons.
- Very small water systems serving fewer than 500 people require about 8 to 10 times as much capital per gallon of water sold as systems serving more than 50,000 persons.

Capital is required in new water source development and raw water transmission, in provision of water treatment plants, and in development of finished water storage and distribution facilities. In the early genesis of small water systems, these capital costs were covered initially by various sources of economic growth that may have included development by a single large industry, by residential real estate developers, or by government grants. The average municipal system is about 60 years old. The average privately owned system is about 30 years old (Dziegielewski, Beck, and Bik 2000). Many of the original capital facilities are now approaching the end of their useful lives and will soon require rehabilitation or replacement. Still disadvantaged by the small scale of the facilities, replacement needs are believed to be about three times as great on a per capita basis in small systems as they are in systems serving more than 3,300 persons (USEPA 1999). In addition, new drinking water standards developed under the SDWA are presenting needs to find new supply sources and/or build more sophisticated new treatment systems.

The combination of upgrading and replacement needs is imposing new capital demands that have not been a major challenge in system finances and operations since these small systems were first constructed. Many small communities face this future with a smaller population base than that which was present during the boom phase that promoted the original development of the system. However, opposite demographic trends are evident in some regions where the outer reach of suburban growth is presenting small systems with new capital demands to expand their systems to serve more customers. While this may help to overcome some of the problems of small-scale operation in the long run, it requires additional capital first to make the transition to larger scale.

Cash is king in the successful operation of a small water system. Researchers have documented a clear correlation between the level of cash reserves in a small water system and the system's ability to meet new regulatory requirements, avoid regulatory violations, respond to emergencies, support training needs, maintain adequate inventories of critical parts and supplies, and maintain equipment in good working order. These relationships are being accentuated with the increasing complexity of regulatory requirements and treatment systems.

The measure of cash flow that has been used in various studies is the ratio of operating revenues to operation and maintenance (O&M) expenses, called the operating ratio. This ratio gives an indication of the extent to which the cash flow of the business provides a buffer over the basic running costs, for contingencies and a generally higher level of preparedness. If this ratio is less than 1.0, it is a clear danger signal that the water system is not taking in enough money to cover all its expenses. Many small systems of all ownership types operate too near this threshold. Often, a bad year or two that produces negative net revenue (operating revenue less O&M expenses) is followed by corrections in rates that brings the water system back to solvency.

In the 25–100 persons served system size category, USEPA data show that 61% of systems have an operating ratio less than 1.0. In the 101–500 persons served category, 43% have an operating ratio less than 1.0. In the 501–3,300 size category, only 25% have an operating ratio less than 1.0 (USEPA 1999). Several researchers have concluded, however, that operating ratios must be maintained at a level no lower than 1.2 to sustain good operational performance by small water systems (Cromwell and Jordan 1999).

As the complexity of water treatment systems and regulatory processes has grown over the last few decades, the challenges of providing adequate operations capability at small scale have grown also. The number of operators per system has remained fairly static, however. On average, systems serving less than 500 persons have about one-and-a-half full-time equivalent positions, whereas systems serving 501 to 3,300 persons have about three full-time equivalent positions. By far, one of the most significant challenges is delivering appropriate training to small systems (NRC 1997). Because small system operators are variously part-time, volunteers, or the lone person fully responsible for the system, it is often difficult for them to break away for a day and attend a training session that may be many hours' drive away from their rural location. Moreover, many training programs are tailored to the needs of medium and large water systems and deal with topics that may have only partial relevance to small systems.

USEPA data show that access to technical assistance differs between systems serving fewer than 500 persons and those serving more than 500 persons (USEPA 1999). More than 90% of public and private systems serving more than 500 persons report receiving some form of technical assistance. The same is true for publicly owned systems serving fewer than 500 persons. However, only 81% of privately owned systems serving 101 to 500 people and 72% of privately owned systems serving 25 to 100 people report receiving some form of technical assistance.

As source water constraints, regulatory demands, or other factors add costs and technical complexity to the day-to-day challenges faced by CWSs, sustainability becomes harder to maintain. Forward-looking and realistic communities thus need to look at new and previously unnecessary and perhaps unpalatable options for addressing the growing challenges to sustainability. Regional solutions provide several avenues that should be considered.

REGIONAL CONSOLIDATION AS ONE POSSIBLE PATH TO SUSTAINABILITY

As noted above, consolidation often is seen as a likely and promising solution for many sustainability problems faced by small and large systems alike. This is especially true where a water system faces a challenge due to regulatory compliance (e.g., where meeting a new MCL would require the adoption of an expensive and complex new treatment technology), or where new source waters need to be tapped (e.g., because region-wide growth has rendered the current source inadequate for meeting local or regional demands, and/or because greater system interdependence provides a hedge against variability in supply availability for any given source).

In these situations, a neighboring system may be able to furnish MCL-compliant or additional source waters, once the utilities are physically interconnected by transmission lines. Or, two neighboring systems facing the same problem may jointly develop a new water source or treatment facility, thereby enabling each system to enjoy greater economies of scale by teaming with its neighbor. In either case, a community water supply system engaged in such a regional solution is likely to save some production expense, enhance water quality, improve supply reliability, tap better technical expertise, or realize some combination of these factors. These are the potential benefits of a regional approach.

On the cost side of the ledger, systems that consolidate on a regional basis will still face some added expense (e.g., for physical interconnections), lose some local autonomy, or both. There also are a host of institutional, legal, regulatory, accounting, engineering, operational, hydrologic, and management aspects that need to be considered and addressed. As detailed in chapter 4, regional solutions entail much more than devising an engineering and/or hydrologic solution to a problem, and it is these multiple other dimensions associated with implementing a regional approach that can pose appreciable concern and create hurdles to what may look like a relatively simple engineering solution.

The basic options noted by NRC (1997) for regional restructuring include:

- Direct transfer of system ownership
- Receivership or regulatory take-over
- Purchase of contract services
- Receipt of technical support on a regular basis

Two general cases of organizational change characterize the most popular examples—contract operations and assimilation into a larger utility system or network of systems. These are discussed below.

When the performance problems facing a small system consist mainly of operational issues, there is a clear indication that contract operations may be beneficial. The deficiencies may arise from a lack of staff training or a more fundamental lack of available labor in the relevant skill areas. Rather than continue to struggle with SDWA violations, service quality complaints, and higher costs of reactive maintenance efforts, it may be more efficient and more cost-effective in the long run to obtain contract O&M services for the system.

Providers of contract O&M services obtain economies of scale by deploying trained staff over multiple facilities within a region and by focusing their business on the specialty of water system operations. Markets for these services have evolved uniquely in every region of the country. There are therefore a variety of providers, but not all of the choices will have evolved in every region. Some of the possibilities for contract O&M providers include:

- Large private companies that focus exclusively on O&M services nationally as well as regionally
- Private companies that own and operate utilities as well as provide contract O&M services within a state under the regulatory supervision of the state public utility commission
- County or regional authorities or special districts established for the purpose of running a network of small systems and/or providing O&M services
- Nearby urban or suburban utilities that have an interest in providing O&M services as an expansion of their mission within the region
- Nonprofit entities such as the state Rural Water Associations that have a specialized focus on small system problems and the flexibility to blend contract support, technical assistance, and training in tailored packages

Like all the options for stand-alone operation of small systems, all of these options for contract support have their share of success stories and horror stories. At the general level, it suffices to say that they are all workable. The best advice is to look for the best fit to the circumstances.

When the signs point to underlying financial weaknesses as well as operational difficulties, the situation may be suitable for assimilation of the small system into a larger entity rather than contract operations. The possibilities include:

- Purchased water or merger deals with nearby small communities to obtain greater economies of scale
- Private companies that own and operate utilities as well as provide contract O&M services within a state under the regulatory supervision of the state public utility commission
- County or regional authorities or special districts established for the purpose of running a network of small systems and/or providing O&M services
- Nearby urban or suburban utilities that have an interest in assimilating small systems providing O&M services as an expansion of their mission within the region

As with the contract O&M options, there are examples of success and failure with all of these assimilation options. There are more differences in terms of ownership issues, however. While contract O&M is reversible, the assimilation of a small systems assets into a larger organization required under these options is a permanent transfer of assets. So, once a partner has been selected and a deal concluded, there is no going back. That need not be a barrier, but just an extra reason to shop carefully.

If a utility chooses to consolidate into some type of regional cooperative arrangement, it will have to relinquish some degree of control over operations to another organization, called a "restructuring agent." The restructuring agent could be the neighboring community's water system, an investor-owned water services company, or some county or regional authority. The NRC notes that a utility should use a business planning approach to assess its options, and the local community should be involved as much as possible to ensure community endorsement (NRC 1997).

Barriers to consolidation of small systems include disputes over who should pay for system improvements, lack of data for assessing what will be involved in assisting a system, requirements that restructuring agents be held liable for violations of drinking water standards by the small system, political resistance to ownership changes, lack of funds to promote feasibility studies, and water resource allocation policies (NRC 1997).

Serious barriers to consolidation also exist for larger systems. For example, a city-owned water system may be a logical partner for (or the core basis of) a more region-wide arrangement for water provision, including not just the central city but also several surrounding communities as well. However, the central city's political leaders may not wish to see "their" water system transformed to some regional entity that includes other political jurisdictions. Likewise, neighboring community leaders may not wish to see their systems subsumed under some larger, city-dominated system.

In addition to matters of "local control," leaders in the various communities may have fiscal reasons for opposing consolidation. For example, the city's mayor may not be willing to lose a reliable source of revenue from the city water department. Large municipal water utilities are often viewed as "cash cows" by the local governing officials—providing a large and reliably stable amount of revenue for city coffers, year in and year out. Likewise, some potential partnership communities may believe that their current water rates are below average and will be replaced by higher region-wide rates if one larger system is formed and then applies a common rate structure across all jurisdictions served. These are among the political and pragmatic realities that often complicate or stall regionalization efforts.

REGIONALIZATION: PANACEA OR LIMITED SOLUTION?

Since the SDWA was first signed into law in 1974, many regulators and other water professionals have been concerned with how small water systems would survive under increased regulation and compliance activities. Early on, it was recognized that some form of regionalization might offer opportunities to efficiently meet many of these challenges. The NRC evaluation reflects this perspective as well (NRC 1997).

In some circles, regionalization was seen as a desirable and perhaps inevitable process through which the multitude of small community water systems would be dramatically reduced in number over time. This consolidation of small systems into larger ones was seen as the inevitable result of regulatory and other cost pressures inducing many small systems to combine in some fashion (either with each other or with neighboring large systems) to take advantage of economies of scale in water treatment and delivery. Indeed, regionalization was seen—and is still seen today by some advocates—as a panacea for the affordability and other compliance-related problems posed to regulators by small water systems (and also the solution to the problems faced by small systems and their customers).

Although there are several advantages to consolidation (as discussed in the following sections), it has not been widely adopted by the U.S. water industry. The U.S. network for supplying drinking water remains highly fragmented (and perhaps increasingly so). The number of small community water systems (CWSs serving 10,000 or fewer people) has been steadily increasing since the early 1960s (USEPA 1995). In 1963 there were approximately 16,700 small water systems (Public Health Service 1965), but by 1994 there were over 54,000 (USEPA 1994). This apparent growth in the number of small systems is probably explained in part by the change in data reporting, permitting requirements, and better record keeping that followed passage of the SDWA. However, the number of small CWSs has remained relatively stable since federal tracking of CWSs improved with the SDWA and its related requirements.

This predominance of small-sized water systems is not the case in some other parts of the world. For example, England and Wales combined have just 10 regional water organizations and 22 water companies providing water and sewer service to 99% of the population of 50 million (Okun 1977). This consolidation grew from a national program implemented during World War II, followed by a regionalization program in 1973 (NRC 1997). The initial consolidation was achieved by encouraging large cities to extend service to outlying small communities or, in areas with no large cities, by encouraging small communities to form regional water boards (NRC 1997).

Consolidation has not occurred at the same scale in the U.S., because of the independent development pattern of the water industry (NRC 1997). Historically, each U.S. town and city developed its own local supply, subject initially to local oversight and then, later, to state and federal oversight (NRC 1997). Small system violations of drinking water standards can be a problem, and 30% of small CWSs violated a water-quality, monitoring, or reporting requirement one or more times between 1992 and 1995 (NRC 1997). While many of the violations involved reporting deficiencies rather than impaired water quality per se, research indicates a strong correlation between monitoring and reporting violations and other serious operational deficiencies in small water systems (Cromwell and Rubin 1995).

While some have gone so far to say that the U.S. water utility industry will most likely be transformed into a few large water companies or systems on the national level within the next decade (Moody's Investors Services 2000), the number of small systems may still be on the rise (it certainly does not seem to be declining rapidly). It is difficult, however, to discern actual trends in the number of small CWSs from the national USEPA databases such as the Federal Reporting Data System (FRDS) and its more recent manifestation as the Safe Drinking Water Information System (SDWIS). This is because some of the variation in the number of systems may be due to improved data reporting and quality control rather than to actual changes in the number of CWSs (USEPA 2002a). In addition, the federal and state databases do not track common system ownership or shared management. Thus, trends in regionalization other than physical interconnection cannot be readily analyzed from the databases (Rubin 2002).

Certainly it is true that the various forms of consolidation can offer many advantages to water systems and their customers. There are numerous examples where some form of consolidation has provided many important benefits, and worked to the general satisfaction of the parties involved. Several of these success stories are provided as case studies later in this report. At the same time, consolidation does not yet appear to be very widespread across the North American landscape, even though more than 30 years have passed since the SDWA was first signed into law. This low-level application of the approach might imply one of two things; either:

1. Consolidation has many real drawbacks that have kept it from more widespread acceptance and implementation
2. Misconceptions about the problems created by consolidation (relative to any benefits to be gained) are so widespread that it has not gotten a fair hearing at the community level.

As in most instances, the truth appears to lie somewhere in between these two extreme perspectives. While consolidation can offer sizable benefits and advantages to many utilities, there also are potential barriers, including technical, economic, institutional, and social constraints. In addition, "regionalization will be successful only when communities feel it is in their best interest and that they will have appropriate access to the utility and influence on its decision-making" (NRC 2002, p. 90). Utilities are thus faced with both challenges and opportunities when considering consolidation options. Water systems and other interested parties need to evaluate objectively the full range of benefits, and also acknowledge and consider the full range of costs, while weighing their options about potential regional approaches.

GENERAL ADVANTAGES AND DISADVANTAGES OF REGIONAL SOLUTIONS

In this section, a generic discussion is provided of the pros and cons of regional consolidation. In the next chapter, the key variations of regional approaches are outlined and discussed. Each approach has its own specific pros and cons, which will be discussed later in this report. However, there are some general observations about pros and cons that can be made about regionalization, broadly defined. These are presented here, to offer a broad overview of the key tradeoffs that tend to be applicable when regional solutions need to be evaluated.

Supply reliability has traditionally been maintained by developing local water sources sufficiently large to meet peak demand under all but the most extreme circumstances. In a growing number of regions, however, the costs and risks of such an approach can be prohibitive, particularly when constraints related to agricultural, environmental, and recreational uses are taken into account. As a result, water supply planners may begin to resist dependence on single sources, and be more likely to consider approaches with multiple water supply assets.

However, consolidation often includes short-run opportunity costs, concerns over loss of local control, increased difficulty of maintaining water quality across sources, and additional implications for infrastructure, delivery, and treatment costs (NRC 2002). Consolidation can meet local resistance, because local utilities (and their management and staff) are often entrenched in time honored approaches and they might be reluctant to embrace change. Local utilities, local governing officials, and the citizens they serve may overly emphasize what they can lose to regionalization, while at the same time they may be unclear of what they could gain.

Coordinating multiple supplies with the individual and combined demands of multiple communities provides a number of challenges to the development of regional plans. Whereas regional supply and treatment facilities may offer the collective advantages of economies of scale and increased supply flexibility, few communities are likely to participate in a regional scheme unless it offers distinct improvements over what each community believes it can achieve individually. Furthermore, even where such advantages may exist, socioeconomic, political, and regulatory issues may be significant obstacles. Incorporating consideration of human resources, engineering, operations, uncertainty, and political, financial, and legal issues into the assessment of these complex and interrelated regional systems is often a complicated process, and methods are needed to illuminate these interrelated issues.

Incentives for and Advantages of Regional Consolidation

Regional consolidation strategies may provide ways for utilities to work together to solve common problems, which would allow for greater financial efficiency, improved reliability, and enhanced quality of service. In addition, regional approaches can offer diversification of supply, which can create benefits in risk management because a single source is more vulnerable to accidents and weather events (e.g., chemical spills, storm water runoff) and security threats (e.g., terrorism). Several types of potential advantages from regional approaches are described below.

Economies of Scale

Spreading fixed capital, operation, and maintenance costs over a larger population will most likely lead to lower per unit costs and subsequent lower rates for customers. Additional efficiencies can be experienced by lower total costs stemming from combined financial, administrative, personnel, and equipment resources.

Increased Financial Opportunities

Access to capital can be a problem for small systems. Consolidation can help in several ways:

- *More Revenues and Assets:* Small water systems have limited revenues and assets, and thus may find it difficult to borrow funds for improvements.
- *Larger Tax Base:* Many small systems are located in rural areas with low populations and low growth. These systems have no predictable larger customer tax base in the future to finance capital improvements.
- *Access to Regional Programs:* Many states give priority to regional efforts when disbursing grants and other forms of financial support.

Elimination of Duplicated Services

Greater efficiencies can be experienced by eliminating any operations or services at one facility if another facility has the capability of an increased load. For example, one utility's billing system may have the capability to serve two systems.

Increased Reliability

Meeting water supply needs with one large source instead of numerous small sources can lead to increased service reliability (and the original set of smaller sources can be retained for back-up or supplemental sources to meet peak demands). For example, in a study of a consolidation approach in the Seattle and Everett water systems, Reese, Palmer, and Nelligan-Doran (2000, p. 8) found that "the current status quo scenario for this region could result in portions of the region struggling to meet water demands while other portions remain relatively water rich. Regionalization alternatives would allow all water customers to have safe and reliable water for many decades into the future."

Increased Flexibility

Because of the greater resources available and the improved economies of scale, regionalization opens up new management possibilities. With more opportunities available, communities can develop strategies tailored to their specific needs and concerns (USEPA 1994).

Enhanced Protection of Public Health

With benefits of economies of scale and potential greater access to state-of-the-art technologies, public health and environmental protection can be enhanced for many jurisdictions. Thirty percent of small CWSs violated microbiological standards or monitoring and reporting requirements one or more times between 1992 and 1995 (NRC 1997).

Skill Improvements

Small systems may not have the revenues necessary to hire technically skilled labor. Being able to offer higher salaries could attract better, more qualified employees. For example, supporters of legislation in Arizona claim that consolidation of school districts could offer higher salaries for teachers in rural school districts (*Peoria Times* 2004).

Service Efficiency

Historically, consolidations in police service departments have brought about shorter response times to public safety requests and ease of work assignments (State of Georgia 2004). The quality of policing can rise under consolidation as a result of more efficient and coordinated use of labor, more flexibility to meet hours of peak demand, enhanced training opportunities, and improved management and supervision (International Association of Chiefs of Police 2003). However, it is not known what efficiencies might be experienced in the water utility service.

Disadvantages of and Barriers to Regionalization

The various barriers to regional cooperation have been relatively well documented in the current literature. Physical interconnection and some forms of nonstructural cooperation can be precluded or made uneconomic by distance (due to the cost of piping and pumping) and by characteristics of the terrain (e.g., rivers or mountains to cross) (Castillo et al. 1997). Legal and institutional factors also arise, such as water rights or environmental permit restrictions on where and how much water can be taken from a river, and where and how much water must be returned to the watershed (e.g., regional approaches to share water resources may result in interbasin transfers that are precluded by law, may cause the loss of water rights, or may create other problems). Other challenges are described below.

Loss of Power and Community Independence

A primary concern of communities considering consolidation often is that it will reduce or eliminate their independence and power to influence their own future. Some smaller jurisdictions might be concerned that they will have less influence on or supervision of a consolidated agency than their larger neighbors.

Differing Management Goals

While neighboring communities share many common needs and concerns, disparities in population, geography, or other characteristics may make it difficult for communities to agree on specific regional projects.

Conflicting Regulations

Regional approaches that cross state or county boundaries may face varying regulations. Jurisdictions would need to resolve issues raised by contradictory or conflicting laws and regulations.

Cost and Benefit Inequities

Cost and benefits of regional projects, although shared, will not necessarily be the same for all communities. Some communities may bear a disproportionate share of costs—relative to the benefits derived—when compared to communities with whom they might consolidate services.

Workforce Reduction

Consolidation often entails employee layoffs, and thus it may be difficult to garner support from officials or citizens. Also, difficulties may arise when personnel from one utility are tasked with increasing service to a larger population.

Issues Regarding Shared Resources

Many municipal utilities share resources—people, construction equipment, telephone and computer systems, etc.—between the water utility and other municipal operations. If there is substantial sharing of resources between the water utility and other government functions, the potential benefits from consolidation may be much less than a preliminary analysis might indicate.

Public Confusion

Sometimes, consolidation leads to confusion over service delivery areas and service providers, but this confusion often settles within a short period of time (State of Georgia 2004).

Debt

It may take many years before a consolidation is able to pay off all pre-existing debts, or at least slow their increase. Communities may not support consolidation if it does not immediately bring about tax savings or fee reductions.

Resistance to Change

No matter how advantageous a regional approach may be, it will reflect a change from how institutions and assets have been managed and owned in the past. Change is hard, no matter what the net improvements might be.

CONCLUSIONS

Water suppliers strive to provide reliable, high quality, and sustainable service to their communities. There are several considerable challenges to this, and regional approaches typically offer advantages that can help utilities realize efficiencies and attain sustainability. Nonetheless, there has been less movement in North America toward consolidation of water systems than has been exhibited in many European countries, indicating that the disadvantages—whether perceived or actual—may create a considerable impediment to the broader use of regional solutions. The balance of this report is aimed at helping provide a balanced and comprehensive assessment of the pros and cons of regional approaches, so that water supply professionals can gain a better and more complete grasp of the wide ranging options available for their consideration, as well as a better and more complete understanding of the pros, cons, and implementation tips that can enhance the chances for success.

CHAPTER 3
AN OVERVIEW OF REGIONAL SOLUTION OPTIONS

This chapter provides a broad overview of the many variations or types of regional approaches that water systems may consider. Also provided is a review of the literature on consolidation, including a review of regional solutions in sectors other than water supply.

OPTIONS FOR REGIONAL APPROACHES

There are numerous options for consolidation efforts. Rather than provide a lengthy text discourse on the options, the research team has compiled useful information and descriptions in two summary tables. Descriptions of various options available for both nonphysical arrangements (e.g., satellite management) and for physical interconnection approaches (e.g., shared source waters and treatment facilities) are provided in Tables 3.1 and 3.2, respectively. The tables also provide examples and describe key pros and cons. The information provided is derived largely from SMC Martin (1983).

In reviewing the material in these tables, it is important to keep several things in perspective. First, there are many different types of regional solutions, and through the years there have been numerous ways in which people have labeled or categorized these options. In this chapter, we begin in the tables that follow with a traditional dichotomy based on the inclusion or exclusion of physical interconnection as part of the approach. However, there are many others ways to categorize and label the options, and later in this report (chapter 5) we move to an alternative construct based on the type of agreements that can potentially be forged between entities. Second, while in this chapter we retain the traditional terms of "pros" and "cons" when evaluating the various options, it becomes more constructive later in this report to focus on the "pros" as benefits that can be realized, and the "cons" as issues that need to be considered and that often can be addressed (i.e., avoided or minimized) if suitable terms can be included in whatever written agreements are forged between the parties to a regional approach.

A REVIEW OF THE LITERATURE

In general, two types of studies on regional consolidation are found in the literature. The first type focuses on the cost and physical barriers that limit the potential for regionalization, especially the costs of physically interconnecting systems. The second type of study examines case studies where regional approaches have been pursued with some measure of success, and describes what led to those successes. The literature associated with each of these types of studies are reviewed below.

Studies on the Cost and Feasibility of Regional Approaches

The cost-oriented studies have a simple basic premise: the infrastructure improvements to make interconnection possible often are expensive, hence physical interconnection may not in the end provide the level of net economic savings that might first be envisioned. This literature includes the following studies:

Table 3.1
Nonphysical options for regional solutions

Option	Brief description	Pros	Cons	Applications of this arrangement	Scenario in which this arrangement would work well
Informal agreement	Voluntary cooperative decision between 2 or more systems to share a commonly needed component.	• Relatively quick and easy to implement • Duration of agreement is adjustable depending on system needs • Provides systems with a means of testing future cooperation • Easy to end	• Not legally enforceable • Easy to terminate • No formal continuity from administrator to administrator	• Sharing laboratory facilities • Storage facilities • Billing equipment • Providing emergency water • Share O&M functions or personnel	• Two neighboring systems are having difficulties obtaining needed resources. They agree to exchange some items (e.g., one system agrees to periodically loan its backhoe to its neighbor, and in return the neighboring system shares some time of its certified operator to assist the other utility). [Note: we do not believe sharing water is an appropriate scenario for an informal agreement; water should not be transferred without a written agreement spelling out legal liabilities, etc.]
Basic service contract	Involves a legal contract between systems, where the recipient retains policymaking and financing control and the provider performs service functions. It is the most common and simplest form of consolidation.	• Easy to create • No restrictions on local autonomy or policy control • No governmental reorganization • Flexible depending on system needs • Economies of scale • Able to provide specialized services not otherwise available • No voter approval required	• Easy to terminate • May be a temporary solution to a long-term problem • Can be expensive if contracting is the only option for a major problem • May require competitive bidding	• Wholesale and retail water purchase contracts • Contracting O&M and emergency services • Water plant O&M • Distribution system maintenance • Billing and collection	• Most commonly used to provide wholesale or retail water when local supplies are of insufficient quality or quantity.

(continued)

Table 3.1 (Continued)

Option	Brief description	Pros	Cons	Applications of this arrangement	Scenario in which this arrangement would work well
Joint service contracts	Sharing or exchange of activities among systems. Generally more complex than a basic service contract, and participants are generally partners rather than contractors. Administrative decisions (planning, contracting, financing, and sometimes operating costs) are made jointly.	• Larger economies of scale • Minimal disruption of administrative structures • More permanent than basic service contracts • More uniform coordination and administration of services • More efficient use of personnel, equipment, and facilities • Able to provide specialized services • Elimination of duplicate facilities • Increase in overall efficiency of service • No voter approval required	• Affects local autonomy and policy control • More difficult to terminate • Benefits to outside jurisdictions that do not compensate participants • Sometimes difficult to distribute overhead and other costs equally between participants • Difficult for participants to provide service themselves if the agreement fails and customers are accustomed to the service already • Increased administrative tasks and costs	• Developing a new water source • Ownership of system facilities (storage, laboratories, maintenance, vehicles) • Purchasing (chemicals, parts) • Exchange or sharing of service activities (O&M, billing and collection)	• Neighboring communities have insufficient or inadequate water supply, and need to develop a new water source.
Regional council of local selected officials	Forum for identifying problems in a region. Encourages common action between participants, to minimize duplicating efforts. Participants are not necessarily bound by council findings or recommendations.	• Centralized planning and coordination • Forum for community and individual input to decision-making • No restrictions on local autonomy or policy control	• Decisions not legally enforceable • No power to raise funds • Relation to other governmental units is strictly advisory	• Evaluating common problems that cross jurisdictional boundaries (e.g., population growth and planning, raw water provision and quality)	• Forming a regional council to review source water quality and quantity, to make recommendations for region-wide source water protection.

Table 3.2
Physical interconnection options for regional consolidation

Option	Brief description	Pros	Cons	Scenario in which this arrangement would work well
Association/ nonprofit water supply corporation	Nonprofit institutions created from consolidated smaller systems to provide water in rural, unincorporated areas.	• Authorized to acquire water sources and construct and operate a water distribution system • May be able to issue bonds secured by assets and revenues • May be authorized to seek federal financing	• No power to tax • Not authorized to issue general obligation bonds • May be under-capitalized • May have difficulty competing with for-profit and governmental utilities for employees • Can be difficult to govern • May not have power of eminent domain	Neighboring small water systems are unable to raise capital or obtain low interest loans to improve treatment or distribution systems. Joining together allows them eligibility for federal financing through Farmers Home Administration grants and loans.
Special districts and authorities (Note: also can be nonphysical)	Units of local or state government that provide one or more services to a particular geographic area.	• Often the only method to provide a badly needed service • May have power of eminent domain • May be authorized to levy special assessments • Can match service areas with service needs • More focused than local government because it is specialized • May have greater financial flexibility than local government • Less restrictive than local government on cooperative agreements • Convenient and inexpensive way to provide service in rural areas • May be initiated by citizens • May not have state-imposed debt ceilings • May be able to pay higher salaries • Provision of service to areas that cross jurisdictional boundaries • Economies of scale	• Bonds may be restricted to revenue bonds, which can be repaid only by user revenues • Powers may be limited directly to those required to provide service • May have limited powers compared to local government units • Susceptible to public opposition because of its permanence • May be difficult to govern • Potential lack of accessibility and accountability • Activities uncoordinated with those of other local governments	Taxpayer or government concern that individual water systems are not able to serve needs of area. Also may be used to avoid Public Utility Commission/Public Service Commission (PUC/PSC) regulation of municipal expansion outside of municipality.

(continued)

Table 3.2 (Continued)

Option	Brief description	Pros	Cons	Scenario in which this arrangement would work well
Annexation	Occurs when a water system extends its service area to include neighboring territory. Will not involve legal annexation of a geographic area if it is a private or nonprofit water system.	• Immediate increase in service area population • Makes use of infrastructure of existing water supply entity • Provision of service to areas outside original jurisdictional boundaries • Annexed area acquires same rights and obligations as rest of service area • Economies of scale • Power of eminent domain • Applicable to other municipal services besides water supply	• Complex implementation • Susceptible to public opposition from those not wishing to be annexed • Voter approval may be required • Can be politically motivated • Not applicable to noncontiguous areas • Capital expense required to service new customers	Unincorporated area adjacent to an incorporated area lacks reliable public water service. Municipal water system needs to either raise rates or extend service in order to remain solvent. Decide to extend public water supply and other municipal services to incorporated area. Incorporated area gets services, expanded service population makes it easier to absorb costs of capital improvements and O&M.

- A study by the USEPA (1993) of small water systems in three states (Alabama, South Dakota, and West Virginia) concluded that as many as 50% of small water systems could potentially engage in some type of external restructuring to reduce the cost of meeting more stringent water quality requirements. The costs of physical interconnection were not fully quantified, however.
- Castillo et al. (1997) developed a study for AwwaRF on restructuring small systems, using geographical location and other data available for water systems. The authors found that physical interconnection between small and larger systems might be economically feasible for up to 35% of small systems in some states, but that in most of the states studied such consolidation would be economically feasible for only 10% to 20% of small systems. Nonphysical consolidation was found to be economically feasible for nearly all small systems.
- On behalf of National Rural Water Association (NRWA), Ottem, Jones, and Raucher (2003) revisited the 1997 study by Castillo et al., and used a more robust Geographic Information Systems (GIS) dataset to analyze the minimum distance that small systems face to connect with nearby larger systems. The Ottem, Jones, and Raucher study made an important distinction between small CWSs in metropolitan areas as contrasted to those located in rural areas of the U.S. In metropolitan areas, on average, small systems are significantly closer to the nearest larger system and serve populations with higher incomes and lower rates of poverty than their rural counterparts. Thus, small rural systems tend to face higher costs for interconnection and have less ability to pay these costs. Additional discussion is provided in a section provided below.
- In another, earlier study for the NRWA, Rubin (2001) also found that small rural systems are, on average, located in communities with significantly lower levels of economic resources than small urban or suburban systems.

In sum, the existing literature acknowledges that regional approaches can hold considerable promise for small systems. However, they also reveal that regional approaches that rely on physical interconnection are likely to be of limited applicability, due primarily to the economic factors associated with the cost of pipeline infrastructure and pumping costs, especially in the context of rural communities with limited economic capacity.

Literature on Case Studies of (Mostly) Successful Consolidation Efforts

The case study literature looks at utilities with experience in regionalization. These case studies tend to detail what factors have affected the success of the efforts.

- Graham et al. (1999) conducted a survey and prepared several case studies to "examine how utility managers and planners can respond to the challenge of growing demand while managing constraints on source development." These case studies looked at a wide array of options that utilities are using for meeting long-term future water demands. A key focus of the report is guidance for utilities on dealing with stakeholders and other utilities within their region, and the authors' recommendations that are displayed in Table 3.3.

Table 3.3
Recommendations from Graham et al. (1999) for managing constraints in order to meet future demand

- Identify key public values and assess their relationship to water supply plan elements
- Emphasize stewardship of water resources
- Use an adaptive management framework to manage uncertainty
- Find common ground between urban and rural communities
- Consider project elements that achieve benefits for other stakeholders
- Coordinate planning among multiple utilities within water supply regions
- Incorporate a 50-year planning horizon in water-resource planning efforts
- Build internal capacity in terms of negotiation, organizational analysis, and political strategy

- More recently, in a report prepared for the USEPA, the Cadmus Group (2002) performed several case studies on efforts to promote system consolidation in several states. The Cadmus Group (2002) looked at six state programs to evaluate what elements of the programs helped promote consolidation, and what elements posed barriers to or created disincentives for consolidation. The six states examined were Alabama, California, Connecticut, Kentucky, South Dakota, and Texas. The finding from the case studies drawn from these states were assembled as a list of approaches, and are summarized in Tables 3.4 and 3.5.

Where the Current Literature is Lacking

The existing literature on consolidation is useful and provides a good start for building an evaluation tool that will help utilities weigh the pros and cons of various regional approaches. However, the current literature is lacking in at least three key areas:

- First, the available studies tend to focus on either monetary costs of physical interconnection or social and political barriers to regionalization—none of the studies identified really tackles both issues.
- Second, the available literature has focused largely on costs of and barriers to regionalization—few studies have illustrated the benefits of regionalization (Reese, Palmer, and Nelligan-Doran 2000), is a partial exception, with the benefits characterized in terms of percent time the source water streams in the region remain above minimum flow targets.
- Finally, the studies have focused largely on detailing past consolidation efforts—few have attempted to provide practical advice to utilities on how they could succeed in future efforts.

In the balance of this report, we aim to address these limitations.

Table 3.4
Approaches to state programs from Cadmus Group (2002) found to create incentives or otherwise help promote use of regional approaches

- Political/stakeholder support for consolidation, with or without legislation
- Legislation mandating or encouraging consolidation for nonviable systems
- Developed policy defining consolidation as a State goal, including long-term plans
- Ability to force nonviable systems into receivership, external supervision, and /or consolidation with a viable system
- Plan review/technical review process required for loan applications or systems modifications
- Higher rankings/additional points given to systems considering consolidation on Drinking Water State Revolving Fund (DWSRF) loan applications
- State requires that consolidation be considered in DWSRF application or other process (e.g., creation of new systems, system expansion)
- Restriction on the creation of new small systems if interconnection to existing system is feasible
- Availability and accessibility of Federal and/or State funding sources
- Development/coordination of funding programs for interconnection transmission lines
- Zero interest loans offered to consolidating systems, under certain conditions
- Loan forgiveness plan
- Using TMF assessments and requirements to compel nonviable systems to consolidate
- Grassroots support for consolidation
- Creation of regional bodies to encourage consolidation and develop regional water plans
- Creation of "exclusive service areas" for existing, viable utilities
- Use of existing sub-State structures (e.g., counties) to encourage and manage consolidation
- Cooperation between State agencies and/or other relevant entities (e.g., Rural Water Association, RUS, Water Board)
- Support for consolidation from the engineering community
- Focus on eliminating individual Non-Transient Non-Community Water Systems (NT-NCWS), especially schools
- Development of new or existing processes to identify consolidation candidates

Table 3.5
Factors in state programs found by Cadmus Group (2002) to create disincentives or barriers to consolidation efforts

- Process places a burden on State financial resources
- Process adds burden to State and utility staff
- State lacks legislative authority to force consolidation
- State lacks data/process to identify consolidation candidates
- Local communities perceive a loss of autonomy and power
- Consumers unwilling to accept rate increases
- Resistance to consolidation by urban growth opponents
- Local communities reluctant to lose water rights
- Larger/viable systems unwilling to take on problems of smaller/less viable systems
- Geography precluding easy/cost effective physical interconnection
- Lack of funding for physical interconnection
- Abundant ground water sources
- Congressional ear-mark grants or grants in general

IMPORTANT DIFFERENCES BETWEEN SMALL COMMUNITY WATER SYSTEMS IN RURAL VERSUS METROPOLITAN AREAS

Emerging evidence indicates that there may be significant differences between small system realities in rural areas as opposed to the circumstances of small CWSs located in metropolitan locations. Table 3.6 presents a summary of national level results from a cursory examination of distance-based consolidation opportunities, from Ottem, Jones, and Raucher (2003). Results for several statistics are reported by THE USEPA-established size category ("very small" systems of 25 to 500 persons served, and "small" CWSs of 501 to 3,300 persons served). These results are broken down according to water systems in urban locations (i.e., within a census-designated "metropolitan area") and rural areas, throughout the U.S.

Because of physical distance, small and very small community water systems are likely to face relatively high costs to connect to larger systems. Over half of small systems are located more than 7.5 miles from the nearest medium, large, or very large system. For very small systems, the median distance is over 6 miles. For both size categories, however, a large number of systems are much farther from larger systems. Small systems and very small systems are on average located approximately 8.5 miles or more from the nearest medium, large, or very large system. More than 25% of systems in both size categories would have to connect to a larger system that is 12 or more miles away.

In addition, the results indicate that there are, in fact, some very noticeable differences between small water systems in rural and metropolitan areas. Since rural systems tend to be located much farther from a larger system than are small systems in urban areas, their costs for any regional approaches relying on physical connections will tend to be much higher. The difference in average and median distance for both the small and very small categories is striking.

- Approximately half of the very small systems in urban areas are less than 4.5 miles from a larger system. In rural areas, fewer than 25% of the systems are within 5 miles.
- For small systems a similar pattern is evident—more than 50% of urban systems are 3.9 miles or less from a larger system, but fewer than 25% of rural systems are within 5.8 miles of a potential partner system.

In conclusion, while regional approaches may provide advantages to CWSs in all areas, the small utilities in rural areas face considerably larger barriers to successfully tapping into regional solutions. Physical distances and lack of financial resources are likely to pose challenges to small rural systems that hope to improve their water service through some form of physical interconnection, unless outside parties can provide financial and other assistance. At the same time, rural systems will most likely have a harder time paying for potential physical interconnections. The household income of the population served by rural small and very small community water systems is generally much lower (more than $8,900 per year less, for both size categories), and the poverty rate is higher (by 2 percentage points or more, for both size categories) than in the population served by small and very small urban systems.

These statistics do not represent every system in every state. On the whole, reliable location data are available for roughly 85% of small and very small water systems. The percentage does vary by state, however. For example, location data are available for all small and very small systems in Connecticut, but are missing for approximately 90% of small and very small systems in Kansas. A similar result is found for medium, large, and very large systems—there are reliable

Table 3.6
Summary of key statistics for small and very small water systems in rural and urban areas*

System type	Sample size	Distance (in miles) to nearest medium, large, or very large facility: 25th percentile	Median	75th percentile	Mean	Population served	Service connections	Household income†	Poverty rate‡
Very small systems (25 to 500 persons served)									
Metro areas	8,406	1.95	4.11	7.25	5.42	147	57	46,706	10.01%
Rural areas	7,975	5.07	9.90	16.33	11.71	160	65	37,770	11.95%
All	16,381	2.87	6.16	11.98	8.48	154	61	42,356	10.95%
Small systems (501 to 3,300 persons served)									
Metro areas	2,271	1.58	3.91	7.39	5.31	1,376	467	46,703	10.04%
Rural areas	3,854	5.75	10.37	15.94	11.74	1,343	515	36,440	12.71%
All	6,125	3.32	7.52	13.28	9.36	1,356	497	40,245	11.72%

* Includes water systems in the following 34 states: Arizona, Arkansas, California, Connecticut, Florida, Georgia, Idaho, Illinois, Indiana, Iowa, Maine, Maryland, Massachusetts, Minnesota, Mississippi, Missouri, Montana, Nebraska, New Hampshire, New Jersey, New Mexico, New York, North Carolina, North Dakota, Ohio, Oregon, Pennsylvania, South Dakota, Utah, Vermont, Virginia, Washington, West Virginia, Wisconsin. Others excluded because of insufficient data.
† Mean value of median household income in all census tracts within 10 miles of water system, weighted by number of households.
‡ Mean percentage of households in all census tracts within 10 miles of water system.

location data for approximately 85% of systems in the U.S., but data are missing for a large percentage of systems in certain states.

ADDITIONAL CASE STUDIES OF SUCCESSFUL REGIONAL CONSOLIDATION

There are additional case studies that are available in the literature that help illustrate some "success stories" from consolidation efforts. In this section, we provide brief summaries of two such cases.

The first case study involves a situation in which the utilities faced a shortfall in source water supply because rapid regional growth spurred demand to levels that the existing source waters could not meet. This is becoming a familiar scenario in many areas of the U.S., where counties that were once fairly rural start to experience a rapid influx of population as the outer reaches of the suburbs expand into their areas.

The second case study involves a situation in which water quality and related regulatory compliance problems prompted consolidation-based solutions for a small system. This is a scenario that is likely to arise with more frequency as new and more stringent regulations are issued over time by the USEPA and state primacy agents.

Regional Approach Motivated by Insufficient Supply

As population grows and suburban areas expand, many rural areas are quickly becoming bedroom communities for larger cities. Oftentimes this rapid growth includes new developments on the outskirts of town and higher density building in town, demanding more connections and more water supplies. Many formerly rural communities are then faced with a very real problem of having inadequate water supplies to service their growing population, which may then limit their economic growth. For many small systems, the prospect of securing a new water supply is daunting and expensive.

In the early 1990s, communities in Christian, Todd, and Logan counties, in Kentucky, were experiencing these same pressures. Communities in these adjacent counties began investigating the problem, and through informal discussions realized that they all faced similar challenges. Leaders in these communities decided that the most effective way of addressing their common challenges was to create a multi-county water commission (Jesperson 2003).

The Logan-Todd Commission represents 12 autonomous water systems, each represented on the commission by one member. Its primary goal was to address the shortage of raw water in the region by finding a suitable new source. Each system retains ownership and operating and maintenance responsibility for its own distribution system. The commission has the authority to build, own, and operate water supply and treatment facilities. It also has the power of eminent domain, but has no taxing authority. The commission sells water at the same wholesale rate to each system, ensuring that all systems pay the same rate.

Once the commission located the best water supply (the Cumberland River in Tennessee), it designed the treatment and distribution system. It was laid out in such a way so water would not have to pass through one system to get to another, minimizing water loss, transmission cost, and capacity utilization issues between member systems (this also may reduce water residence time and hence, limit disinfection byproduct formation). The system design process was led by plant operators from member systems who advised engineers on particular system needs (Jesperson 2003).

The treatment plant was ready for service in 2003. Highlights of this major undertaking, as an example of successful regionalization, include:

- Voluntary regionalization was achieved by recognizing common needs.
- Communities maintained autonomy over their own systems.
- Overall cost to consumers was less than it would have been, given other alternatives.

Table 3.7 provides summaries of additional case studies where consolidation has been used to address shortfalls in source water supplies.

Regional Approach Motivated by Regulatory Noncompliance

Many small systems are challenged by poor water quality that causes them to violate regulations and that potentially may pose health risks and damage consumer confidence. For many systems with problems posed by one or more contaminants, the cost of installing new treatment equipment can be very high and would result in a substantial rate increase for customers, surpass the utility's access to capital, and challenge the technical know-how of the utility staff. Some systems have found consolidation to be a feasible option for solving major noncompliance issues.

The City of Panora, Iowa, is one such system. Panora is approximately 45 miles from Des Moines, with 700 mostly residential connections. Panora had chronic seasonal problems with nitrates in their finished water: nitrate levels would exceed the MCL by 20% to 40% during the spring and early summer. They also lacked the financial resources to install new treatment equipment and train their Grade 2 certified operator for Grade 3 certification (USEPA 2002b).

To address these problems, Panora chose to pursue informal cooperation with two nearby systems. To address their water quality issue, they created an arrangement with the nearby community of Lake Panorama, which did not have the same nitrate issues. They found that connection with and purchase from Lake Panorama was the most cost-effective option over the long term. They decided to connect and draw on Lake Panorama's water only during high nitrate times, when they blended the low-nitrate water with their high-nitrate water to comply with the MCL. This option allowed Panora to cost-effectively comply with the nitrate MCL, thereby protecting public health and improving consumer confidence.

To address their problem with retaining a Grade 3 certified operator, Panora looked to nearby Des Moines Water Works. Des Moines Water Works operators mentored Panora's operators in the Grade 3 certification process, and evaluated the possibility of remote monitoring by Des Moines. This arrangement has saved Panora money in the short term through lower training costs, and a remote monitoring arrangement could save them substantial money in the long term (USEPA 2002b).

Finally, this form of consolidation has not resulted in any loss of assets or local autonomy for Panora. Panora has simply uncovered means of working cooperatively with their neighbors to achieve their water quality objectives and attain compliance. Table 3.8 offers summaries of other case studies in which consolidation has been used to address chronic problems with water quality and regulatory compliance.

Table 3.7
Regional solution examples for system facing supply and distribution challenges

City	Number of connections	Pop. served	Technical issues	Managerial issues	Financial issues	Consolidation arrangement	Specific actions
Clarion, Pa.	u*	203	• Distribution system needed upgrades—leaks and malfunctioning meters • Did not serve all of community, and already purchasing water from a larger, neighboring entity	• Managed by a board of part-time volunteers, unable to deal with technical and financial difficulties • In general, a poorly run, under-capitalized small system	• Were purchasing water from large private system PA American Water (PAWC), and could no longer meet financial obligations to them • Rates were 20% higher than PAWC's state-wide prices	Ownership transfer (privatization)	• Sold system to PAWC for amount of debt owed to PAWC • Obtained low-interest loan from DWSRF for Pennsylvania to expand connections to households with inadequate wells • Assigned part-time certified operator to oversee system
Sully, Iowa	u	841	• Need backup well and additional storage capacity • 26% water loss in system • Low flows in case of fire • Need higher capacity treatment system for iron and radium			Connection partnership with nearby system Central Iowa Water Authority (CIWA)	• Full service bulk connection with CIWA • Sully retains control of distribution and retail services • Less expensive than doing it themselves

(continued)

Table 3.7 (Continued)

City	Number of connections	Pop. served	Technical issues	Managerial issues	Financial issues	Consolidation arrangement	Specific actions
Guildhall, Vt. and Northumberland, N.H.	u	u	• Northumberland had drilled wells unsuccessfully • Residents objected to treating river water for new supply	• In this instance (but not in all cases) the use of interstate water sales might have required inter-municipal agreement and Congressional approval (Note: Congressional approval probably not typically required for water sales across state lines)		Purchase agreement	• Rural Development intervened and encouraged Guildhall to continue purchasing water from Northumberland • As a funder, Rural Development was able to convince them to stay together • Guildhall and Northumberland were able to negotiate inter-municipal purchase agreement and obtain Congressional approval for this interstate water purchase
Todd and Christian counties, Ky.	u	45,000	• Inadequate supply • Same problem affecting numerous adjacent communities			Regionalization	• Created regional commission tasked with operating supply and treatment facilities • Each individual system still in charge of O&M for own distribution system • Each system sets own rates • Commission takes on no debt for members

*u = unknown

Table 3.8
Regional consolidation examples for systems facing compliance and monitoring challenges

City	Number of connections	Pop. served	Technical issues	Managerial issues	Financial issues	Consolidation arrangement	Specific actions
City of Panora, Iowa	700	1,175	• Finished water exceeds nitrate MCL • Grade 2 operator needs to get Grade 3 certification	• Difficulty retaining operators	• Inadequate financial resources to pay for operator certification and install nitrate treatment	Informal cooperation	• Connected to and purchased water from Lake Panorama's low-nitrate source water, for use during high nitrate times • Evaluated and are pursuing remote monitoring by Des Moines Water Works • Des Moines operators mentoring Panora's operator in Grade 3 certification
Sanford (Lee County Water Plant), N.C.	60	149	• Lack expertise to make improvements • Lack regular supervision by qualified operator • Only periodic technical assistance		• Inadequate financial resources to hire qualified operator	Contractual assistance	• Entered into management, operation, and maintenance contract with private firm • Firm hired and supervises plant employees, including part-time qualified operator • Firm does financial management and billing through central office
Aurora, S.D.	250	500	• Finished water exceeded nitrate MCL • Nearby wells had high nitrate levels • Plant operator inadequately certified and also had non-plant responsibilities		• Inadequate financial resources for nitrate treatment	Contractual assistance	• Aurora and Brookings shared cost of pipeline to interconnect systems • Aurora customers pay slightly more for water now, but less than if nitrate treatment had been installed • Aurora uses old well for fire dept.
Pittsfield, N.H.	620 (75% res.)	1,860	• Transmission and distribution systems oversized • Frequent main breaks • In violation of Surface Water Treatment Rule; facing big fines • Part-time operator insufficient for proper system oversight and customer communication	• Insufficient communi-cation with customers	• Could not secure low-interest loan, so burden of new treatment plant was very high	Ownership transfer (acquisition and satellite operation)	• Nearby, larger utility with substantial technical, managerial, financial resources purchased system • New owner obtained State grant and refinanced loan to complete new treatment plant • New owner received a Housing and Urban Development (HUD) Community Development Block Grant to upgrade distribution system and improve water quality

PUBLIC OPINIONS ABOUT REGIONALIZATION: LESSONS FROM OTHER SECTORS

Water supply provision is not the only service sector in which consolidation has been advocated, considered, and at times applied. Police and fire protection, primary and secondary (K–12) education, and library services are among the traditional local government service offerings for which consolidation has been debated and, in some instances, adopted.

Polls in New Jersey, for example, have revealed that a large segment of the general public is inherently hesitant to support consolidation of local services such as schools, libraries, police and fire protection, senior activities, and road maintenance, including snow removal (Eagleton Institute of Politics 2004). While a majority of New Jersey residents (58%) had heard nothing at all about regionalization, 49% support regionalization of local services, and 39% oppose it. Roughly half of surveyed state residents believe regionalization will save money and lower property taxes, and only one in five (19%) think it will have a significant impact. Residents are somewhat more likely to think that regionalization will worsen the quality of local services (27%) rather than improve them (19%), but most (49%) think it will have no impact. Residents in New Jersey's developing towns (58%) were somewhat more likely than those in stable suburbs (43%) and urban areas (39%) of the state to favor the regionalizing local services to hold down taxes.

Similar results were experienced in a comparable Eagleton poll in New Jersey from 1998, showing that public opinion on this topic has not changed much over the past 6 years (Eagleton Institute of Politics 1998). The 1998 poll also asked additional questions concerning how the public thinks about the regionalization of specific services. The public services for which the public most favors consolidation of are senior citizen activities (83%), libraries (82%), and road maintenance and snow removal (69%). The public was less supportive of consolidation of fire protection (61%), police protection (54%), and schools (52%). However, opposition to consolidation of all of these services grew significantly when consolidation entailed impacts not in the respondents' favor:

- Sharing schools with a neighboring town was favored by 52% of those surveyed, while 40% oppose. Not surprisingly, support is stronger when the school system is seen to be "trading up" rather than down in terms of wealth. However, in a separate question, 68% prefer to have their own local schools if given the choice, while 23% prefer mandatory regionalization; and, if a reduction in property taxes accompanied mandatory regionalization, 51% preferred local schools, while 41% preferred regionalization.
- Sharing public libraries with neighboring towns was favored by 82%, while 14% oppose. However, 49% opposed consolidating library services if it meant their town's library would be smaller, while 44% favor.
- Sharing police protection with a neighboring town was favored by 54%, while 41% oppose. However, 53% oppose consolidation of police forces if it meant firing the police chief in their town, while 39% favor.
- Combining departments for road and street maintenance and snow removal with another town was favored by 69%. However, 60% favor combining departments for road and street maintenance and snow removal with another town, if it was under the mayoral oversight of their own town, while 34% oppose. And 49% favor combining departments for road and street maintenance and snow removal with another town if it was under the mayoral oversight of a different town, while 44% oppose.

- Sharing fire protection with a neighboring town was favored by 61%, while 36% oppose. However, 69% oppose combining fire departments if it meant it might take a little longer for fire-fighters to get to a fire, and 28% favor.

The public opinions noted above for a range of services traditionally provided by local government (or quasi government) agencies confirm some basic observations often made regarding water service. In general, whether for police, fire, library, or water service provision, people tend to favor local provision of essential services or, if shared, to have the service more under control of *their* locality than their neighbors. A broad desire for maintaining a sense of local control appears to be a consideration that needs to remain in mind when regional approaches are explored, regardless of what type of service is being considered.

CONCLUSIONS

There are many ways of categorizing regional water service options, and the literature often has made a distinction between physical interconnection and options that do not entail physical links between water systems. Regardless of how the options are labeled and categorized, there are a broad array of possible choices, and each tends to offer distinct advantages and disadvantages, depending on the circumstances a given utility faces.

The literature reveals some good examples of how and where regional approaches have been beneficial to water utilities, and also reveals how regional approaches have been viewed by the public for provision of local services other than water. The key for water agencies and other interested parties is to identify which regional options (if any) appear best suited for their specific needs, and then to forge an agreement with their neighbors so that the benefits can be realized and potential concerns addressed. The chapters that follow are developed to help guide water agencies and other concerned parties through this process, so that they can better understand their options, narrow down the range of options to those approaches that are likely to be best suited to their circumstances, and strive to form an agreement of the type that will most likely generate benefits and address their concerns.

CHAPTER 4
KEY ELEMENTS OF REGIONAL SOLUTIONS

This chapter discusses the broad suite of key issues or "elements" associated with most regional solution approaches. This provides an overview of the many aspects of regional solutions that need to be recognized and addressed, looking beyond the engineering and hydraulics considerations that often are the subject of technically focused discussions of regionalization.

The key elements examined here include considerations such as who owns what assets, who is obligated to pay for what, and who has what legal liabilities and regulatory accountabilities. Also discussed are matters associated with local autonomy and control, distribution system impacts on water quality, hydrologic (e.g., water resource and water rights) issues, energy cost concerns, security, and disaster preparedness and response.

The intent is to provide readers with a broad perspective of the myriad elements they need to consider in regional approaches, so that they can be better prepared to recognize and address the wide-ranging implications of consolidation options when they become engaged in evaluating or pursuing regional solutions. The topics introduced in this chapter can be viewed as a "checklist" of items that a water agency should consider as it contemplates whether it wants to form or enter a regional approach. For water systems that are already a part of a regional approach, this checklist can be used as a self evaluation guide, to help identify areas in which they may wish to expand, update, or reform their existing agreements.

In addition, most if not all of these elements can (and should) be addressed as part of any regional partnership agreement or contract, since these agreements provide the basis for anticipating and rectifying the many issues that may arise in a regional solution context. These agreements are discussed in greater detail in the next chapter.

ENGINEERING AND HYDRAULICS

In the past, regionalization often was conceived of as the physical interconnection of two or more systems, such that waters from one source and treatment plant could be transmitted to one or more neighboring communities for distribution to the customers of the adjoining system(s). In such a context, there are fairly standard engineering and related hydraulic issues to resolve. These are associated with sizing and locating pipes and pumps to effectively move the necessary volume of water over the applicable distance and terrain, at the desired pressure.

While the design and operation of inter-CWS water transmission networks is not always simple or inexpensive (Castillo et al. 1997), the nature of the challenge is fairly well understood, and the technical expertise to address these issues is widely available. Therefore, this report does not dwell on this relatively technical and widely recognized aspect of regional solutions. Instead, through the sections below, the focus is on institutional and related issues that might otherwise be overlooked or under appreciated by water professionals and other interested parties.

LOCAL AUTONOMY

A core issue associated with regional approaches is local control. This is a factor that needs to be recognized as a potential "make or break" element in many (if not virtually all) efforts to pursue regional approaches. Local governments and communities tend to have a strong interest in retaining control over their own destiny. Regional approaches that might in some manner curtail local autonomy, or create the perception that some local power may be lost, are likely to have appreciable hurdles to surmount.

There are several reasons why local autonomy may become a critical concern in deliberations over regional approaches. Among these reasons is the sense of local *identity* that may be linked to water supply provision. Most towns traditionally have developed with their "own" local service provision network (e.g., local water, local schools, local law enforcement), and the prospect of in some way turning over those traditionally local assets or functions to a neighboring community or a larger entity (such as a county-level authority) may be seen as eroding the local flavor that differentiates the home community from its neighbors.

A perceived or real loss of local *control* over core services is a key obstacle that must be considered. Residents recognize that water (and wastewater) service provision is essential to the community's survival and well-being, and therefore are reluctant to turn over the reins and/or assets to some "outside" entity. This concern is heightened because water and wastewater service provision frequently is used as the *de facto* land use planning tool by localities, especially in areas experiencing growth. The location and timing of where and when a community agrees to provide water-related services along its outskirts typically is what determines what (and whose) local lands get developed (e.g., transferred from farms or woodlands into residential subdivisions). This phenomenon of "zoning by infrastructure" is a common practice for managing or promoting growth, and is so fundamental to the traditional functioning of many communities that any potential loss of governing capacity in that regard is not likely to be relinquished easily.

In addition, water and wastewater operations can be an important source of cash flow, personnel, and equipment for a municipality. A local government may view these resources as being important not only for routine operations, but also as a way for the community to provide services in emergencies (for example, having construction equipment that can clear debris after a flood, or having a pool of personnel with commercial driver's licenses that can operate trucks or other heavy equipment during emergencies).

Accordingly, one of our key findings from reviewing the literature and conducting case studies is that a regional approach to water supply provision is most likely to be acceptable to the parties (and their constituents or customers) if the solution is structured in a manner that enables local communities to retain control over key assets, authorities, and responsibilities and, hence, retain more local control. Local ownership of assets (e.g., the local distribution system and any future extensions, as well as related equipment), and retention of local customers (and associated revenues) are two ingredients that enable a local community to retain at least some degree of control over many critical aspects of local decisions and operations. Regional approaches that help assure continued local autonomy over key issues—such as the rate, location, and scale of local growth—can be the core factor that renders a regional approach palatable or not.

It also should be noted that regional approaches can become a key mechanism to assure greater local autonomy, as opposed to being a threat to (or otherwise limiting) local control. Regional solutions inevitably involve written agreements and contracts between neighboring entities. These agreements can serve as a viable basis for ensuring that localities retain ownership or control over key elements that might otherwise be diminished through other means. For example, communities in the shadow of a large city can forge regional agreements on water provision that preserve their independence and forestall potential annexation by their large urban neighbor. Indeed, this form of protection from a large city was a key motivation for forming the world's largest regional provider, the Metropolitan Water District of Southern California (MWD), as discussed in chapter 7 (which contains our case studies).

WATER QUALITY CHANGES: IMPROVEMENT OR DEGRADATION

One potential advantage of a regional approach is the possibility of delivering customers higher quality water than they would otherwise receive from their local system. This may arise, for example, where higher quality source water is furnished from the regional supplier, or where the regional supplier can furnish a higher level of treatment than the local provider. It is this potential for higher quality water that has sparked much of the past interest in regionalization, especially from a regulatory compliance perspective.

However, a regional approach can also give rise to tap water quality problems and regulatory compliance challenges. Of particular concern is the potential for higher levels of disinfection byproduct (DBP) formation as associated with treated waters having longer residence times in regionally extended distribution networks. Simply put, the larger the transmission network, the longer the time for byproducts to form between the point of disinfection and the consumers' tap. The potential for elevated DBPs raises concerns about public health protection, because DBP exposure may be associated with elevated risks of some cancers and, perhaps, an increased risk of adverse reproductive outcomes (e.g., miscarriage) (USEPA 2003). Further, because DBP levels in distributed waters are regulated by the USEPA under the SDWA, regional strategies that link utilities into "consecutive systems" can pose challenges and complications for regulatory compliance (see below).

Concern over potential water quality degradation extends beyond DBP formation. Longer distribution networks and finished water residence times, and increasing the number of pressure zones and pumping locations, increase opportunities for infiltration of low quality (possibly tainted) waters into the transmission lines. Likewise, maintaining adequate and compliant residual disinfection levels in larger distribution systems also can be a challenge—increasing the potential for microbial risks or regulatory compliance problems, or both.

Water quality concerns in physically interconnected systems extend beyond health risks and assciated compliance challenges, and also include aesthetic issues such as taste, odor, appearance, and pressure. In physically connected systems, these problems can arise from blending waters that have different pH or chemical characteristics because they stem from different sources or are produced from different treatment regimes. Also, interconnections that reverse the direction of flow in water transmission mains can scour materials from the pipe wall and degrade the quality of water delivered to customers.

The bottom line is that when systems physically interconnect, they need to be alert to potential water quality implications that may arise as unintended consequences. Utilities will also need to be prepared to address these water quality implications.

In addition, a simple change in source water or disinfectant, as may arise where one agency reaches an agreement to deliver wholesale water to another utility, can spark customer complaints. These complaints may simply reflect customers experiencing "different" water from what they had grown accustomed to, rather than necessarily the receipt of lesser quality water. Nonetheless, customer complaints about water quality should be anticipated, and often will require management attention in a regional water sharing arrangement.

REGULATORY COMPLIANCE AND ACCOUNTABILITY

As noted above, water quality issues can arise—for better or worse—when water systems physically interconnect. This creates challenges for regulatory compliance. One specific regulatory issue of concern pertains to compliance with the Stage 2 DBP rule (as finalized January 2006) in what are referred to as "consecutive systems" (USEPA 2003, 2006).

A consecutive system is one in which one (or more) utility retails water that it receives wholesale from a regional provider. More precisely, the USEPA defines consecutive systems as "public water systems that buy or otherwise receive some or all of their finished water from another public water system" (USEPA 2006, p. 392). In the DBP context, the regulatory framework for the Stage 2 rule stresses compliance at various key points in the retailer system's distribution system (to limit DBP levels faced by people who receive that water at their taps). Thus, the point of compliance is within the retailing water system, meaning that the retail agency is accountable for regulatory compliance.

However, in consecutive systems, the retail CWS receives its water from a regional provider, and this wholesaler is the entity that has control over most of the critical resources and practices (e.g., source water, treatment processes, residence times) that determine DBP levels and, hence, the compliance status for the retailer. In many instances, there are limited practical or feasible opportunities for the retailing agency to impact water quality (e.g., the systems' configurations have evolved in such a way that there are few if any opportunities for water-receiving systems to add treatment or otherwise manage key water quality parameters in a technically feasible or reasonably cost-effective manner). Further, the wholesale-retail arrangement probably was adopted, at least in part, so that the retailer would not need to have water treatment expertise.

In other words, in consecutive systems, there is likely to be a disconnect between the entity that is legally accountable for regulatory compliance (i.e., the retailer), and the entity that has considerable expertise in, and control over, the relevant water quality factors (i.e., the regional wholesale provider controls much of what impacts DBP levels near the retail customers' taps). This type of disconnect—in which the entity with accountability is not the same entity that has primary expertise and control—can create considerable compliance-related problems.

Clearly, regulatory and other systems work best when responsibility and control are fully aligned (rather than disconnected), so that the controlling entity is also the party accountable for compliance. This alignment of accountability with control is the key to properly incentivizing entities into acting responsibly.

Perhaps the best way to address this sort of disconnect in regional approaches is through the contractual agreements through which wholesalers and retailers set the terms and conditions of their relationship. A contract (or other legal agreement) entered by both parties can, if properly written, effectively provide the mechanism for aligning control of water quality with regulatory compliance accountability. Thus, when water systems consider entering into regional approaches, they need to (1) be alert to existing and pending issues of regulatory compliance, and (2) develop written agreements that do not leave them wholly responsible for the actions taken by other parties (i.e., shift some compliance responsibilities to those entities with more control over DBP formation). One option is to have contractually set water quality parameters (i.e., DBP levels) measured at the customer turnout. This could entail wholesalers providing additional regional or local turnout treatments. Another option is for the wholesale customer to provide treatment (e.g., partial granular activated carbon, GAC, treatment) at the turnout. This can be for aesthetic or compliance reasons.

FISCAL RESPONSIBILITY AND CAPABILITIES

When water systems consider entering into regional arrangements, they need to be alert to how the terms of the agreement can impact their financial circumstances. For example, the agreement should specify clearly which entity pays (or receives compensation) for the assets and services involved, the basis for establishing the amounts to be paid or received, and the course of action (or consequences) should one party fail to fulfill its financial or service obligations. In concert with these terms, water systems should explicitly address which entity will own and control which assets, and who will be responsible for any past debts that may be linked to the assets or services in the agreement.

In addition to the issues above—which are the types of terms that are routinely addressed in any good contractual agreement—water agencies and other interested parties should also consider several other factors that are somewhat unique to water supply providers. For example, water systems may be eligible for various forms of financial assistance. These include grants or loans from federal agencies, such as issued under the USDA's rural assistance program, or USEPA's State Revolving Fund (SRF) loans that carry favorable terms. In many states, access to SRF or other monies are impacted by the ownership of the entities seeking support (e.g., in some states, privately owned systems, or investor-owned for profit systems, may not be eligible for some programs or receive lower priority).

Therefore, when water agencies consider entering a regional approach, they need to be alert to, and fully consider, all the potential financial implications of the options before them. As in any sound business dealings, it is important for entities to perform a comprehensive due diligence and develop a written agreement that provides for suitable controls and protections. The degree to which these issues arise will depend on the type of regional approach being considered, with greater attention required where there is a potential merging of entity assets from one system into another, or where multiple systems expect to share some or all assets with another.

MANAGEMENT AUTHORITY

In concert with financial, regulatory, and other issues of accountability and responsibility, it is important to understand how a regional approach would define which persons and entities will hold authority for making what decisions. As noted throughout this chapter, the key for water systems is to think ahead about what management authorities arise in the regional context, and then consider how these authorities might best be assigned. This entails defining who is authorized to make what decisions, and what terms and boundaries may be defined for the types of decisions involved (e.g., dollar spending limits that the designated manager is not authorized to exceed without explicit consent from other entities). The best approach to addressing this issue is to make sure management authorities, and related terms, are suitably and explicitly established in whatever agreements a water system tries to forge with its regional neighbors. This can pertain to forging new agreements, or to revisiting and renegotiating (or renewing) existing contracts with neighboring systems.

PUBLIC AND PRIVATE OWNERSHIP

In some instances, a regional approach may involve some sharing or transition of assets between one or more publicly-owned utility and one or more privately-owned system. While there is nothing inherently problematic with either public or private ownership, or with private- public partnerships, the mixing or transfer of system ownership types can pose some challenges. For example, eligibility for certain types of advantageous funding (e.g., government grants or SRF loans) may be limited to publicly-owned entities in some states. Or, some local stakeholders may have strong philosophical distaste for one or the other form of water supply ownership.

Moreover, in many states, private ownership would require approval, and continuing oversight, from the PSC/PUC. While PSC/PUC regulation can be beneficial for both utilities and consumers in many circumstances, it also can result in additional costs, as well as delays in undertaking certain types of projects.

As noted throughout this chapter, the key is for water systems and stakeholders to be alert to possible ownership-type implications (advantages or drawbacks), and be prepared to address them if they wish to proceed.

LINKS TO OTHER LOCAL SERVICES

One subtle and often important aspect of local water service provision—especially in municipally-owned systems—is the joint use of some capital or labor assets across two or more public entities that provide different local services. For example, a town water department may well share assets such as trucks, backhoes, and select labor skills with a sister municipal agency such as the town roads department. This sort of informal cooperative arrangement can be important in municipalities. Also important can be local jobs (e.g., a skilled water department employee or manager) that might be threatened under a regional approach.

Unless such local asset sharing and employment arrangements are recognized and accommodated in some fashion, they may create hurdles (or concerns) when regional water supply options are being considered. If consolidation of a town water system into a regional entity may remove access to a backhoe or skilled labor often shared with the town roads department or other local government service providers, then a key to developing an acceptable regional solution in water may entail some recognition and accommodation for the non-water local agencies.

WATER RESOURCE AND ENVIRONMENTAL ISSUES

One key motive for regional approaches is they can provide an opportunity to better align the scale of regional water resource availability with the sum of the water demands generated across a region by multiple water entities. Significant advantages often can be derived by matching regional water resource supplies to regional water demands.

However, in some instances, local hydrologic conditions are such that regionally-based water resource extraction can create problems with regard to water rights and interbasin transfers. For example, it may make good sense in many ways to explore a regional approach that taps a single source for the region-wide supply, but if water rights from that single source are not available to meet the regional demand, then the approach may not be viable (although some creative alternatives, such as water rights transfers, options, or exchanges, may be available).

Another important concern arises when the locations from which water is extracted differ from the places or water bodies into which the associated wastewater is discharged. In such circumstances, the lack of return flows can create severe problems in the source water body, including instances where "permanent" streams and rivers run dry in some months in regions where surface water flows would otherwise run year round (e.g., the Ipswich River, in Massachusetts, in summer months when water extraction demands peak; see Raucher et al. 2005).

This phenomenon has arisen in many locations because of federal policies to encourage regionalization of wastewater collection, treatment, and discharge during the 1970s and early 1980s. Federal grants for wastewater system construction and upgrades promoted relatively large regional approaches for sewer collection systems with large centralized treatment plants and discharge. In many areas, this approach consolidated wastewater from many towns and localities, but these towns typically had their own water supplies that often were drawn from different sources. The consolidated wastewater typically is centrally treated and discharged to a single receiving body that often is in a different hydrologic basin than where much of the water originated as source water. With continued growth, the various communities extract more water over time. Because little of the extracted water finds its way back to the original source (or does so a considerable distance downstream), this can produce extreme low flow conditions in some locations.

In such circumstances, there could be an immediate hydrologic benefit from a regional water supply approach that matches water supply extraction location and scale with the general location and scale of treated wastewater discharge. This can help keep more return flows in the same basin from which the waters were extracted. In turn, this may provide environmental and other benefits by solving low flow problems in small water bodies.

As an interesting side point, there currently is a growing interest in exploring *decentralized* approaches to wastewater treatment (e.g., the Water Environment Research Foundation, WERF, has several federally-funded projects on this topic). One benefit touted with decentralized wastewater control is that it can re-align discharge and recharge back to the local basins from which the source waters are extracted.

In summary, there is good reason to recognize the hydrologic implications of the scale and source of water supply across a hydrologic region. Planners should consider the implications of having the scale and source of water supply differ from the scale and receiving body of wastewater in the region.

ENERGY COSTS AND RELIABILITY

Energy is a critical input to pumping and treating water, and often is a significant portion of the total operating expense incurred by a water supply agency. The average water utility from a recent benchmarking study spends 11% of its operating budget on energy (AwwaRF 2003), but for some utilities energy expenses amount to over 35% of operating outlays. In California, for example, energy costs are on average 34% of utility expenses (CEC 2005). Collectively, water and wastewater utilities are the second largest user of electricity in the U.S., using about 3% of all electricity in the country (Beecher and Rubin 2000). Further, these energy expenses incurred by utilities may only reflect a fraction of the total energy use embedded in water supply provision, especially where waters are imported across long distances and changes in elevation and thus require considerable pumping.

The water sector's dependence on and level of energy use is likely to increase in the years ahead. This increased energy use is anticipated because utilities are expected to increase applications of relatively energy intensive treatment technologies, including ultraviolet (UV) light disinfection, ozonation, and reverse osmosis and other membrane processes. This transition to high energy-using processes is developing in response to several factors. These pressures include regulatory demands (e.g., to reduce DBP formation, while enhancing microbial inactivation), demands by consumers for better tap water aesthetics, and the need to transform low quality source waters (including saline coastal and groundwaters, and recycled water) into sources suitable for potable or other uses (e.g., industrial processes or irrigation).

As energy prices continue to rise (often well ahead of the general rate of inflation), and as energy supply reliability becomes increasingly suspect (e.g., due to problems with the power grid, limited electricity generation and fuel oil refinery capacity, reliance on imports), water suppliers need to become more focused on managing their energy demands and addressing their increased vulnerabilities to power disruptions and price increases.

One potential drawback to large, physically interconnected regional water systems is that they often require pumping large volumes of water over considerable distances. The cost and reliability issues associated with the associated energy demands need to be considered in these circumstances.

On the other hand, a relatively large regional water entity may have some energy-related advantages. For example, as a large single buyer (or as a group of smaller systems that join together in a joint purchasing arrangement), a regional entity may have greater leverage to secure relatively favorable terms with power suppliers. Also, because water is often a key input to energy production (e.g., the large-scale need for cooling water), a regional approach may provide water systems with some advantages in negotiating water sales to local energy production facilities.

The key point is that when water utilities consider any regional approach, due consideration should be focused on the energy implications. Utilities should focus on both their energy demands as well as on their options for securing their energy supplies. Issues of energy costs and energy reliability (and of the energy-related risks posed to water agencies due to uncertainties about future energy prices and reliability) should be an important part of the evaluation of a potential or existing regional approach.

SECURITY AND DISASTER PREPAREDNESS

In the post 9-11 and post Hurricane Katrina world, greater attention is being paid to how water utilities recognize their vulnerabilities and takes steps to address them (e.g., enhance security, develop back-ups for critical assets, and train for emergency response). These factors should be part of the evaluation for a regional approach to water supply.

For example, numerous independent water suppliers in a region can share information and resources in vulnerability assessments, and in developing emergency response plans and mutual aid agreements (e.g., see Herrick et al. 2006). Water systems that are physically interconnected (or considering doing so) also need to consider the disaster implications of a larger, connected system. There can be advantages and disadvantages to being part of a larger system in the event of a terrorist attack or a natural disaster such as an earthquake. For example, a larger system that retains access to multiple source waters has a more diversified portfolio of supply options. This diversity of supply may increase vulnerabilities in some ways (e.g., more potential resources to protect), but it also can provide more resources with which to respond to an emergency (e.g., alternative water sources are available in the event that one is lost).

In contrast, if a regional approach entails shutting down various localized sources and consolidates its systems onto a single source, this in effect reduces the region's portfolio of water supply options. Thus, it may be advantageous to retain connections to diverse localized water sources, as a source of insurance against disaster, even if a regional approach is designed to rely primarily on a single larger source or treatment facility.

EMERGENCY RESPONSE AND SERVICE RECOVERY

A critical facet of dealing with a disaster (whether man-made or natural), is the ability to continue to provide water (business continuity) and to rapidly restore service (service recovery) in those cases in which delivery is disrupted. This is critical for public safety purposes such as fire protection and basic sanitation, as well for human health (having potable water available to people, even if using temporary measures such as bottled or trucked water provision). When a regional approach to water supply is being considered or used, one key issue to take into consideration is whether and how the regional approach will address business continuity and service restoration issues in the aftermath of a disaster.

One generally overlooked aspect of this issue is how utilities will set and operationalize priorities for restoring service. As a practical matter, service cannot be restored immediately to all users or communities; and service recovery efforts needs to focus on critical needs first. While the prioritization of service restoration is a very important issue with respect to any local retail water supplier and the customers it serves within its own retail service boundaries, the issue becomes increasingly complex for a regional system where there is a primary water source and/or utility that serves as the principal water provider for multiple systems and jurisdictions. Regional supply arrangements can create challenges and complications because they include multiple political entities (i.e., a collection of politically independent towns or cities). In the aftermath of a disaster that curtails service across the region, there is a potential for debate, grandstanding, and finger-pointing in the midst of an already challenging disaster response, as the elected officials in each locality may clamor if their constituents were not among the first to have service restored.

Thus, it is important that regional approaches be considered in light of the service restoration issues that may arise in a disaster response. A regional approach can provide a larger entity with more resources to make rapid and large-scale repairs, but even a large regional provider can be stretched thin when multiple jurisdictions simultaneously need and demand attention. As in the other issues described in this chapter, it is important to recognize these issues in advance, and to try to establish suitable terms into contracts or other agreements forged in any regional arrangement.

ARTICULATING THE BENEFITS (AND COSTS)

There are many potential benefits to be realized by regional approaches, especially if the most suitable type of regional arrangement is identified and pursued, and if a written agreement is forged that addresses the relevant issues in a sound and constructive manner. However, one potentially large obstacle may be the inability of the entities to estimate and clearly articulate the benefits that are anticipated. This can leave local communities and stakeholders with an incomplete and perhaps unbalanced perspective. Local citizens and agencies tend to focus on the costs or potential detriments of a regional approach, but may have a harder time recognizing the benefits they can expect in return. It will thus be important (yet challenging) to prepare a careful and objective analysis to help identify, estimate, and effectively communicate the benefits that are likely to be realized.

CONCLUSIONS

There are numerous elements to be evaluated when a water utility is considering a regional approach to water supply provision. The degree to which some of these elements may arise will depend on several circumstances, such as the type of regional approach being contemplated (e.g., informal sharing agreements raise fewer of these issues than would a formal merging or transfer of assets), as well as local conditions (e.g., whether water rights issues and/or interbasin transfers are involved).

In all instances, however, the same key theme emerges. Water systems and other stakeholders need to:

1. Be alert to identifying and considering a wide range of potential implications
2. Perform a suitable amount of due diligence so that the relevant issues are identified and considered
3. Try to address all relevant issues through a written agreement that provides suitable degrees and assignments of control and authority, so that the parties to the agreement are appropriately protected (i.e., have their concerns and issues managed) while simultaneously enabling the parties to realize the important benefits that the regional solution can offer

The key is thus to recognize and forge the right types of agreements. This applies to new regional arrangements that are being contemplated, and also to existing regional approaches in which in-place contracts and agreements may need to be modified to reflect changing circumstances, emerging issues, and increased awareness of the broader suite of implications and opportunities afforded by the regional alignment. The next chapter is aimed at assisting agencies in this critical aspect of forging suitable types of agreements for regional solutions.

CHAPTER 5
A TYPOLOGY OF REGIONAL SOLUTIONS

As the literature reviewed in chapter 3 makes clear, the possibilities for regional collaboration are very numerous and inventive. This can make it difficult to obtain an initial orientation on the subject. It makes it a challenge to decide where to start in considering all the possibilities. It is therefore useful to re-summarize this extensive range of ideas and develop a simplified structure that can help to make initial distinctions between the options.

After some sifting and sorting, it is our finding that there are six generic types of regional collaboration. These are distinguished from each other along a gradient extending from sharing and collaboration at one end of the spectrum, to consolidated ownership and operations at the other extreme. Our typology of regional solutions consists of the following:

- Mutual Aid Arrangements
- Sharing Arrangements
- Water Purchase Arrangements
- Collaborative Water Resource Development
- Contract Services Arrangements
- Consolidation

The essential differences among the categories are anchored in what the participants are committing to do together. As a best practice, we observe that these commitments are best documented in written agreements even at the sharing end of the spectrum, where informal collaboration may already exist. A written agreement provides a number of benefits, including:

- It confirms mutual commitments and therefore provides clarity and confidence in the relationship.
- It provides a means of confirming that the "win/win" benefits envisioned by the parties are understood and that there is a commitment to deliver those benefits.
- It provides a framework within which to explicitly identify, address and ameliorate a wide range of issues and concerns that may arise in consideration of the intended collaboration. The very process of negotiating a written agreement inaugurates a process of resolving issues and concerns.
- It provides a possible vehicle for expanding the possibilities for further collaboration down the road, by initiating a practice of negotiating collaborative agreements between partnering utilities. Ultimately, recognizing that an agreement is a control instrument can go a long way toward alleviating concerns regarding who is in control of collaborative arrangements.
- It offers a process (i.e., negotiating an agreement) that should expose the problems, pitfalls, and concerns that may arise in a regional collaboration. If these issues can be resolved through negotiations and by making provisions to address them in the agreement, then they are much less likely to present problems later on and can enable all participants to proceed with greater confidence. If issues and concerns cannot be resolved through the process of negotiating an agreement, it is a sure sign that the deal being discussed is not a viable collaboration.

A key point is that negotiations leading up to a potential regional collaboration agreement provide a means of testing the proposed collaboration. There is benefit even when negotiations fail since it confirms that a beneficial win/win opportunity has not been missed, but was not really as beneficial as it seemed and was not the right choice. Negotiations also advance a dialogue between neighbors that may produce alternative ideas for collaboration or improve the possibilities for collaboration down the road as circumstances change.

In this chapter, we review each of the six generic types of regional solutions with the goal of identifying the essential elements that should be covered in developing collaborative agreements. We begin with a description of each type of agreement, and a characterization of the common forms that are observed. We then discuss two of the key elements that an agreement should address: the benefits to be delivered and the issues/concerns to be addressed/ameliorated.

This framework—"benefits and issues"—is offered as a replacement for the familiar "pros and cons" framework that is often used to characterize the set of factors that enter into a decision on whether to pursue a regional collaboration. We offer "benefits and issues" as a suggested new paradigm for decision making. Whereas "pros and cons" is a static assessment of current status, "benefits and issues" points towards the logical next steps—at the negotiating table. It brings the key points together that are critical to deciding whether to adopt a regional solution:

1. There must be compelling win/win benefits that provide rewards to all the parties.
2. There must be complete and satisfactory treatment provided for all the issues and concerns raised by each of the parties.
3. The best way to guarantee that benefits will be provided as desired and that issues and concerns will be treated satisfactorily is to fully articulate these aspects of the deal in the process of negotiating a formal agreement to guide the proposed collaboration. Negotiation is the acid test that is required to apply the "benefits and issues" paradigm.

As discussed elsewhere in this report, our decision making paradigm of "benefits and issues" also casts a new meaning on the concept of "success and failure" in regional collaboration. In our view, any process that enables neighboring jurisdictions to discuss potential regional collaborations and articulate their desires in terms of "benefits and issues" is a success regardless of whether a regional collaboration results from these discussions. If the discussion of "benefits and issues" does not lead to agreement, it is clear that the opportunity is not ripe for regional collaboration. Such a discussion is a success because a clearly inferior option has been acknowledged as such and set aside while, at the same time, the neighbors have developed a better appreciation of each others' needs that might lead to more promising opportunities at some point farther down the road.

MUTUAL AID ARRANGEMENTS

Mutual aid agreements are an important first step in regional collaboration, involving arrangements for mutual assistance during emergency events.

In the source of supply and treatment areas, the simplest mutual aid between neighboring water systems consists of emergency interconnections that enable temporary provision of treated water to prevent or mitigate customer outages in the event of treatment plant failure, drought emergency, source water contamination episodes and other major outages.

In the distribution system operation and maintenance area, the most common type of mutual aid between neighboring water systems might involve lending repair crews and specialized equipment, and providing specialized items of spare parts inventory to help restore service after a catastrophic event such as an earthquake or flood.

Closely allied forms of mutual aid might extend the relationship to providing technical assistance in the event of treatment emergencies or in the diagnosis and resolution of distribution system contamination events. This might involve the loaning of staff with special skills and specialized laboratory capabilities.

Benefits

Reliability is an extremely important attribute of water supply. Although the public is somewhat accepting of the fact that emergencies happen, the public also expects utilities to recover as quickly and as capably as possible. This helps to restore and reinforce confidence in utility staff and in the utility organization.

Backstopping your own capabilities with those of neighboring water utilities can add a significant buffering to your ability to endure adverse events. Moreover, it can be made available at very little cost compared to what it would cost you to provide such "stand-by" or "backup" capabilities in any other way.

Mutual aid arrangements promote good neighborly relations that can provide a foundation at both the political level and the utility level for additional forms of mutual collaboration. Mutual aid responsibilities provide a good professional challenge and opportunity for staff that can lead to broader professional relationships, benefiting all in the region.

Potential Issues

There is broad public and political acceptance of the benefits of being a helpful neighbor during emergencies. Implementation of mutual aid arrangements are therefore more capable of breaking free of the "us" and "them" context that creates a barrier to many other types of inter-jurisdictional transactions.

When one or more of the neighboring systems is of a different ownership type, the ability to provide mutual aid should not be restricted as long as a straightforward process of cost reimbursement is followed. This should ease any concerns of regulators or governing Boards regarding cost recovery. If one or more of the systems is regulated by the state PSC/PUC, the commission may need to approve the agreement before it can take effect.

There may be concern over the potential liability exposure of "Good Samaritan" activities undertaken by a water system's staff in another water system's jurisdiction during the course of an emergency. This risk can be covered by agreement and made consistent with "Good Samaritan" laws in your state.

SHARING ARRANGEMENTS

Sharing arrangements between neighboring water systems can cover a broad range of facilities and services.

In the source of supply and treatment areas, an obvious opportunity exists to obtain economies of scale through cooperative bulk purchasing of treatment chemicals and other supplies. In states that permit electricity to be purchased from competitive suppliers, joint purchasing of electricity may result in lower costs.

Another obvious opportunity to economize exists in the sharing of specialized resources that are made more affordable when several water systems share the cost. Such resources may include: specially equipped vehicles and construction equipment as well as highly skilled distribution system operations staff such as crews trained in: line repair work, line flushing and rehabilitation, line replacement, valve maintenance, leak detection, pump maintenance, and tank inspection and maintenance. It could also include highly skilled treatment plant operations staff, technical library collections, pilot plants, laboratory staff and lab facilities. For example, a modern gas chromatography mass spectrometry (GCMS) used to scan for trace contaminants might be beyond the means of a single water system, but quite affordable to a collective. The partners may even share similar interests in using such devices in wellhead or watershed protection.

In the management and administration area, a typical example might be the sharing of financial bookkeeping and billing/collections staff. In circumstances where water system operations staff are fully occupied with operations, or simply not trained in administration, it may make sense to borrow support for these functions from a nearby partner that has extra capacity of this type.

Specialized resources are often less than 100% utilized by an individual utility, creating a win-win opportunity to share extra capacity. Sharing offsets the cost of extra capacity for one party while reducing the cost to the other party who might not otherwise be able to afford it.

Benefits

Joint purchasing can benefit all water systems partnering in such arrangements, both large and small. All partners benefit from reduced prices of key inputs like treatment chemicals and electricity. In states that permit electricity to be purchased from competitive suppliers, joint purchasing of electricity may offer an important example where joint purchasing could reduce costs.

Benefits from sharing specialized resources and facilities arise from the increased capacity to afford a higher level of expertise, equipment, and processes. Sharing of specialized staff resources can enable small water systems to afford needed expertise when their needs, or their means, are not sufficient to warrant hiring such personnel. Larger systems benefit from sharing such staff resources as a means of offsetting some portion of their cost of maintaining such expertise and as a means of providing growth opportunities for such staff.

Sharing arrangements promote good neighborly relations that can provide a foundation at both the political level and the utility level for additional forms of collaboration.

Potential Issues

Where it is allowed, joint purchasing of electricity can be a very involved process. The issue of reliability is as important as price. The best deal may affect how a utility has to manage its electricity demand patterns (load characteristics). The special needs of individual utilities should not be allowed to get lost in the attempt to devise a group transaction.

Joint purchasing of bulk treatment chemicals conflicts with existing patterns of relationships with product representatives who often provide informal technical assistance to systems as part of their overall service offering. Difficulties could arise if potential co-op partners have different allegiances. As in other areas, the lowest price may not always be the best deal; product quality and delivery logistics are also key.

Involvement of staff from another entity in financial matters such as bookkeeping and billing/collections may raise issues of local control. It must be stressed that these are only services and not managerial tasks.

A wide range of things could change and affect the balance in a sharing arrangement, but they are all adjustable within the context of a working agreement that can be adapted to correct any flaws. The basis for cost sharing and availability must be understood by all the parties and adapted as needed.

Joint ownership of a shared resource may be prevented by public vs. private differences in the institutional structures of the participating systems. However, it should be possible to accomplish sharing through an agreement where the party with the largest capacity requirement assumes ownership with appropriate commitments from the other parties to support a share of the fixed costs for the unused capacity. Sharing of capital assets obtained with bond proceeds may require confirmation that no bond covenants prevent the rental of unutilized capacity.

Sharing of personnel may raise issues for employees covered by collective bargaining agreements unless job performance requirements are similar enough in the partnering utilities to be considered equivalent. Existing agreements must be searched for any other specific issues that would have to be resolved. Additional issues may be raised if personnel sharing is used as a means of implementing downsizing of a utility's staff.

Other potential issues to be considered in sharing arrangements include:

- If one or more of the systems is regulated by the state PSC/PUC, the commission may need to approve the agreement before it can take effect
- Sharing agreements may raise issues regarding insurance coverage for affected personnel and equipment
- Limitations on the liability of the partners involved in sharing also should be covered by agreement

WATER PURCHASE ARRANGEMENTS

Water purchase agreements are a very common type of collaborative arrangement between neighboring water systems. These may be one-on-one arrangements between two systems. In large metropolitan areas, they may be broader regional collaborations that offer multiple suburban jurisdictions the opportunity to purchase wholesale water from centralized facilities on equivalent terms.

The purchase of treated (i.e., finished) water relieves the purchaser of the need to develop and protect water sources; as well as the need to construct, operate and maintain treatment facilities.

Delivered water is required to meet SDWA standards at the point of transfer, but compliance with standards measured in the distribution system such as those involving lead, coliforms and disinfection by-products are the responsibility of the purchaser.

Because only water and money change hands in such an agreement, ownership of facilities is completely independent among the parties. However, because the seller will build and own more capacity in order to provide the additional quantity, the seller often will require a long-term contract with the purchaser.

Treated water purchases should be governed by written contracts between buyer and seller. If one (or more) of the utilities is regulated by a state PSC/PUC, the contract may be subject to commission approval. Purchase water agreements should specify at a minimum:

- The term of the agreement and any options or conditions for renewal.
- The price of delivered water in addition to any provision for escalation over the term of the agreement. Prices may vary for average day versus peak season deliveries. In particular, the level of guarantee provided for peak day delivery may have a large influence on price.
- Delivery specifications including quality (SDWA compliance) and quantity. Quantity delivered should be defined in terms of both peak period flows and average day flows and line pressures at the point of transfer. The method of measuring flows to document deliveries also should be specified.
- Remedies available to both parties should either party breach the agreement.
- Emergency operating procedures in the event of drought, treatment disruption, or other emergencies.

Benefits

Whenever a community faces significant costs for source development and/or treatment facilities and yet is close to a neighboring system that is better situated to produce water economically, a water purchase arrangement may offer win-win benefits for the two partners. Buyer and seller can both benefit from establishing long-term contracts that enable the seller to undertake source development and treatment projects more economically at larger scale.

Many times, jurisdictional boundaries are so out of sync with topography in a boundary area that purchasing treated water from the neighbor on the other side of the boundary is an obvious least cost method of providing water service in the boundary area versus the cost of extending a community's own system.

Sometimes, having an alternate source of supply to add redundancy to a system will significantly improve overall reliability of service.

Potential Issues

Distance and topography are key factors; water cannot be transported very far or very much against grade without compromising any economic advantages in production. Moreover, recent SDWA standards present new performance challenges in extended distribution networks.

Pricing is a deciding issue in the viability of water purchase agreements. If buyers are reluctant to recognize sellers' needs for capital investments (including a return on capital), or if sellers are unwilling to accept a reasonable premium in the interest of making collaboration attractive, then it may be difficult to get the two sides together on pricing.

When historical water purchase agreements come due for renewal, there is an opportunity to negotiate new relationships to realize additional win-win benefits for all parties. However, both sides have to recognize how the world has changed and not let their impressions of pricing be anchored in the past when current and future costs may reflect different circumstances.

Buyers may feel that committing to long term agreements may cause them to lose some control over the fate of their water supply, but the other party's financial commitment to expanded capacity is the balancing commitment that should provide assurance for the term of the agreement. Over the longer term, it is the hope that a trusting relationship will evolve to make both parties fully comfortable in these commitments.

Often the community that has the water is not the same as the community that has the growth and development. In such circumstances, both can profit from a water purchase arrangement since the benefits of regional economic growth may spill over jurisdictional boundaries. It is reasonable to set aside some capacity to support local growth, but it is most beneficial to be realistic about growth patterns as indicated by market forces.

COLLABORATIVE WATER RESOURCE DEVELOPMENT

An entity is created to coordinate the planning, development and operation of supply and/or treatment facilities at a regional level.

The approach recognizes that because water resources are part of a natural system, the opportunity to manage them in the most beneficial way is greatest when they are managed as a total system. This is facilitated when the boundaries of the management entity match natural system boundaries as closely as possible.

Various levels of coordination are possible within a collaborative regional organization. River basin and major aquifer planning organizations exist in many places to coordinate planning and data collection within the natural system's boundaries. A next level of coordination involves collaborative operation of water withdrawal and storage facilities by multiple water providers in order to obtain the benefits of optimizing the overall system—generally, increased yields and drought reliability. Sometimes these organizations are formed to help comply with water rights limitations or government-mandated water allocations.

One step beyond operational coordination of existing water withdrawal and treatment facilities is collaboration in the development and operation of new supply and treatment facilities. In these arrangements, only the new facilities are jointly owned and operated by the regional entity.

A further level of regional integration is represented by wholesale water suppliers that own and operate most or all of the supply and treatment facilities. In these arrangements, individual jurisdictions own and operate their distribution systems, but purchase all of their water from the regional entity.

Regional organizations for collaborative planning, development and operation of water supply and treatment facilities are of three broad forms:

- Interstate commissions that involve watersheds spanning more than one state. Each state and the U.S. Congress must approve the terms of the interstate compact that governs the commission. Typically, each affected state has a representative on the commission's board.
- State chartered or mandated organizations that are defined in specific state legislation that requires their formation. Board representation usually consists of involved jurisdictions and may also include state-appointed representatives.
- Organizations created under general state laws involving intergovernmental cooperation are usually termed water districts or water authorities. The formation of these entities usually takes place through the consent of the participating jurisdictions

following procedures established in state law. Board representation usually consists of the partnering jurisdictions.

In some states, a county government or regional council of governments may assume the role of a collaborative regional entity for coordination of water resources development.

Where privately owned water systems are interspersed with public ones within the boundaries of a water resource region, the private entities usually cannot be formal partners in these public collaborations, but they may be able to voluntarily participate in planning efforts and voluntarily commit to support common strategies for resource management.

Benefits

Depending on the size of the geographic area being served, water resource development and treatment projects can be more economical with increasing scale. Win-win benefits may be available to all participants in a collaborative regional effort to plan, build and operate such facilities as all parties benefit from their share of the cost savings.

In addition to economies of scale, there also may be economies of scope. By coordinating the development and use of water sources across a region, it is possible to "optimize" the overall functioning of facilities by operating the components as part of an interactive and adaptive regional system. Such optimization strategies can:

- Enhance overall supply reliability during drought periods by allowing water to be withdrawn from the best mix of sources to promote overall system performance under adverse operating conditions.
- Enhance overall supply reliability and quality at lower cost due to the flexibility to draw upon multiple supply sources during periods of unfavorable treatability (salt intrusion, high turbidity, etc.) or in the event of contamination incidents.
- Enhance the sustainability of the overall regional supply strategy by moderating the production demands (e.g., drawdown levels) imposed on any individual source and moderating the environmental stresses imposed on any individual source (e.g., preserving in-stream flows during critical periods to support ecological needs).

The larger the area covered, the greater these benefits are likely to be, but the benefits of centralizing water production may need to be balanced against increased costs for transporting the water over greater distances. Additional economies of scale and scope in the source protection area can be realized when the regional network encompasses the entire watershed or wellhead protection areas of the sources involved.

Potential Issues

- Loss of control. Smaller entities within the region may be concerned that their special needs might be ignored by a large, regional organization.
- Organizational change. The new organizational entity may be viewed as a threat to the existing organizations within each partnering jurisdiction. One way around this issue may be to define the new entity as having only a coordinating role with respect to existing facilities, leaving control in the hands of existing owners and operators. This

may also help to lessen issues over loss of control and legal/financial barriers to the transfer of assets to the new entity.

- Haves vs. have-nots. When some jurisdictions have rights to more water than they need, while others face reverse conditions, it can be difficult to develop a method of compensation that is fair to all involved. This can be especially difficult in areas where older communities have water surpluses and declining tax bases because of business or industrial closings, while outlying communities have increasing tax bases but need water because of growth in housing and businesses.
- Growth vs. no-growth. A large scale effort to provide water supply can be misinterpreted as a pro-growth initiative because it is large and ultimately driven by demand projections that may embody growth. It is possible to craft collaborative organizational designs that can mitigate growth concerns by embracing environmental sustainability as an operating objective and working to encourage demand management (conservation) as a condition of supply expansion.
- Funding. In a number of instances, the stimulus for creation of such regional entities was provided by a state mandate. In some cases, states have created such entities directly and provided funding for the entity's operations. In other instances, states have provided grants to facilitate the start-up process. Other states offer grants or loans in support of construction or conservation projects involving regional collaboration. Where state, county, or regional funding is not available, costs of the regional entity typically would be provided by each participating jurisdiction.

CONTRACT SERVICE ARRANGEMENTS

Contract service agreements provide a means of outsourcing the operation and maintenance of source of supply and treatment facilities for specified periods of time. Under such arrangements, the public or private water system continues to be the owner of the water supply and treatment facilities. Only the operation and maintenance function is outsourced.

In many cases, the water system continues to own and operate the distribution system with its own staff. However, it is also possible to outsource the distribution system O&M function using the same type of contract service agreement.

Numerous private vendors will compete to provide O&M services. The same services may also be obtained from non-profit organizations such as the Rural Water Association or state agencies in some places. In addition, some nearby large water systems and engineering firms may also offer to expand their staff resources and compete with service firms to assist neighboring water systems in a contract capacity.

There is frequently a need to emphasize operational priorities over administrative priorities in small water systems. Contract provision of administrative and management functions can help to fill this void. Contract service agreements can provide a means of assuring the efficacy and cost-effectiveness of necessary functions such as billing and collections.

Despite the choices available in the marketplace for such services, geographic coverage of the different providers may limit the field of competitors and the number of bidders. As a result, it is important to evaluate alternative offers carefully in terms of both price and quality/reliability. Business references for the persons actually proposed to do the work are especially important.

Service contracts typically have a short term (1–5 years) in small systems; larger systems may have longer terms of up to 20 years. The longer term arrangements often involve capital investment by the service provider which could produce cost savings over the long term that are to be shared on a pre-agreed basis with the system owner. The basis for termination as well as options for contract renewal must also be specified.

Contracts must define the transition process from current operation by water system staff to contract operations. This may involve the transfer of water system employees to the contractor's payroll. The complete terms of such transfers must be specified.

Since the cost of operations and maintenance depends to some degree on the condition of the assets—and since the water system remains the owner of the assets—it is necessary to specify the condition in which the assets are to be maintained, the means and frequency for monitoring asset condition, and the differentiation between maintenance expenditures that are the responsibility of the contractor versus capital replacement or refurbishment expenditures that remain the responsibility of the owner.

Although exact methods of operation and maintenance are the domain of the contractor, it is typical for the owner to specify boundary conditions in terms of conformance with the treatment plant O&M manual as well as strict compliance with all state and federal regulatory requirements. Other performance requirements may be specified relating to service quality and reliability.

Contract provisions should also cover administration and coordination between owner and contractor, including owner access to facilities, security, record keeping and reporting, and emergency operating protocols.

If the agreement provides for certain costs to be paid by the owner (for example, some agreements may require the owner to pay electricity costs), the agreement must specify the basis for documenting any such costs.

Benefits

Water system operation and maintenance requirements are becoming increasingly sophisticated. With a limited supply of qualified staff in the labor market, it is becoming increasingly expensive and difficult for small water systems to obtain and retain the skilled labor they need to perform these functions. Outsourcing via contract service agreements may provide a means of using market forces to keep pace with these trends.

Contract service agreements can provide a means of controlling the costs of operating distribution system facilities. While it may not always be wise to select the very lowest bidder, the competitive process may provide a constraint on costs. Although service agreements may have to allow for such things as escalation of energy costs, they can provide a structured arrangement within which costs can be controlled.

Properly structured contract service agreements should provide a very high level of assurance against even minor violations of SDWA and Occupational Safety and Health Administration (OSHA) standards. Violations are typically stated as grounds for terminating a service contract, providing a strong incentive to the contractor.

Potential Issues

Any use of outsourcing raises understandable concerns among the existing workforce. In the case of water treatment plant operators, it is possible that employment with a contract services firm might provide a better career opportunity than the existing water system can offer. Reputable service firms maintain a commitment to training, provide access to broader corporate resources, and offer an opportunity to work in a variety of treatment plants. Transfer of employees must be carefully deliberated with employees, labor representatives, and proposed contractors.

There may be concern about loss of control over water treatment. Clear explanation of contract provisions and safeguards can help to allay these fears and make plain that the contract is a control instrument.

Performance incentives and savings can erode over time. This can be a particular concern with long-term contracts where the contractor is freed from re-competition until a far-off point in the future. It also can be an issue with short-term contracts which frequently have renewal clauses that make contract renewal almost automatic and seemingly removed from competitive pressure. To address these concerns, it is necessary to first assure that the initial competitive process is conducted in a manner that will bring forth the best offers in terms of price and quality/reliability. Following the initial competitive result, it is then incumbent on the owner to negotiate and manage the contract in a manner that maintains a reasonable balance of emphasis between cost control and service quality.

Outsourcing O&M or administrative functions is not the same as selling the water system. The system owner will need to carefully monitor the performance of the contractor; approve and finance capital expenditures; review requests for contract modifications, change orders, and force majeure declarations; ensure that preventive maintenance is occurring in accordance with sound industry practices; and otherwise take actions to ensure that the contractor is abiding by the terms of the contract and not allowing the utility's assets to deteriorate.

CONSOLIDATION

Ownership and operation of source of supply and treatment facilities is merged under the auspices of a single organization. It may or may not also involve merging storage and distribution facilities of the constituent water systems.

This could involve creation of a larger entity for the purpose of consolidating two or more smaller systems or it could be the absorption of a smaller system by a larger one. When ownership and operation of source of supply, treatment and/or distribution system facilities is merged under the auspices of a single organization, it is likely to also involve merging management and administration.

The approach recognizes that there are inherent economies of scale in supply development, treatment, and O&M. This includes not only capital cost savings in rehabilitation or expansion of facilities, but also operating cost savings due to bulk procurement of supplies and due to the needs for specialized staff expertise and specialized resources such as laboratories.

In addition to cost savings, a larger entity may be more capable of accessing the capital financing needed to sustain, enhance and expand supply and treatment facilities and to rehabilitate, replace, enhance and expand distribution facilities.

There may be advantages from the joint operation of a supply strategy that is comprised of multiple source waters having differing characteristics with regard to reliability of yields and environmental sustainability.

If the consolidation is limited to supply and treatment facilities, the merged entity would become the wholesale supplier to the constituent systems that would each own and operate its own distribution systems.

The consolidated entity could be a water district or authority created under provisions of applicable state laws. Under this approach, the consolidated entity typically would include representation from the constituent jurisdictions.

If one utility absorbs the assets of a neighboring system to form a larger utility, the transaction may come under rate regulation by the state PSC/PUC, depending on the specific requirements of state law.

Private investor-owned utilities can also serve as a consolidating entity for bringing dispersed supply and treatment facilities under common ownership and management. Such arrangements usually are regulated by state PSCs/PUCs.

Cooperatives and other private, non-profit forms of organization also can be appropriate vehicles for achieving regional consolidation.

Benefits

Depending on the size of the geographic area being served, water resource development and treatment projects can be more economical with increasing scale. Win-win benefits may be available to all participants in a collaborative regional effort to plan, build and operate such facilities as all parties benefit from their share of the cost savings.

In addition to economies of scale, there also may be economies of scope. By coordinating the development and use of water sources across a region, it is possible to "optimize" the overall functioning of facilities by operating the components as part of an interactive and adaptive regional system. Such optimization strategies can enhance:

- Overall supply reliability during drought periods by allowing water to be withdrawn from the best mix of sources to promote overall system performance under adverse operating conditions.
- Overall supply reliability and quality at lower cost due to the flexibility to draw upon multiple supply sources during periods of unfavorable treatability (salt intrusion, high turbidity, etc.) or in the event of contamination incidents.
- The sustainability of the overall regional supply strategy by moderating the production demands (e.g., drawdown levels) imposed on any individual source and moderating the environmental stresses imposed on any individual source (e.g., preserving in-stream flows during critical periods to support ecological needs).

The larger the system boundary, the greater these benefits are likely to be, but the benefits of centralizing water production may need to be balanced against increased costs for transporting the water over greater distances. Additional economies of scale and scope in the source protection area can be realized when the system boundary encompasses the entire watershed or wellhead protection areas of the sources involved.

Distribution system operation and maintenance is becoming an increasingly complex and sophisticated task that is more favorably accomplished by a larger scale organization. This results from:

- Increasingly complex SDWA regulations requiring ever more sophisticated attention to monitoring, reporting and recordkeeping.
- Increasingly complex interrelationships between distribution system operating regimes and SDWA compliance as measured at the tap.

- More demanding requirements for worker health and safety in distribution system operations.
- New needs to assess and address security threats to distribution system assets.

Distribution system assets are aging and in need of capital reinvestment for rehabilitation and replacement. Larger utilities are likely to have better access to capital than smaller ones. In many situations, there is additional demand for system expansion to serve new growth that presents new capital requirements.

The management and administration function of water systems is becoming an increasingly complex and sophisticated task that is more favorably accomplished by a larger scale organization. Especially when assets are merged or consolidated under common ownership by a larger entity, it is necessary to consolidate management functions from an efficiency perspective.

Potential Issues

- Loss of control. Smaller entities within the region may be concerned that their special needs might be ignored by a large, regional organization.
- Asset Transfers. The value of the assets to be transferred to the new entity must be established using market assessment techniques. Sometimes, there is reluctance to accept the assessed value. State PSCs/PUCs may also impose different valuation approaches. Finally, the ability to transfer assets may be constrained by provisions of outstanding loans or by the requirements of grants that were used to finance the assets.
- Organizational change. The new organizational entity may be viewed as a threat to the existing organizations within each partnering jurisdiction. Especially if cost savings are to result from more economical staffing levels, measures will have to be taken to address workforce impacts.
- Haves vs. have-nots. When some jurisdictions have rights to more water than they need, while others face reverse conditions, it can be difficult to develop a method of compensation that is fair to all involved. This can be especially difficult in areas where older communities have water surpluses and declining tax bases because of business or industrial closings, while outlying communities have increasing tax bases but need water because of growth in housing and businesses.
- Growth vs. no-growth. A large scale effort to provide water supply can be misinterpreted as a pro-growth initiative because it is large and ultimately driven by demand projections that may embody growth. It is possible to craft organizational designs that can mitigate growth concerns by embracing environmental sustainability as an operating objective and working to encourage demand management (conservation) as a condition of supply expansion.
- Funding. In a number of instances, the stimulus for creation of regional entities was provided by a state mandate. In some cases, states have created such entities directly and provided funding for the entity's operations. In other instances, states have provided grants to facilitate the start-up process. Other states offer grants or loans in support of construction projects involving regional collaboration. Where state, county, or regional funding is not available, costs of the regional entity typically would be provided by each participating jurisdiction.

CONCLUSIONS

One of the central challenges inherent in the consideration of regional collaborations among water systems is the potentially expansive range of possibilities to be considered. The fact that there is an expansive array of potential regional approaches often can become an impediment to the consideration of options for regional collaboration. If the options are too numerous to comprehend readily, it takes too much effort to overcome this difficulty in a public forum. When an idea with multiple dimensions and options is put in play in the public sphere, it is likely that different groups will develop different views. This can result in confusion and frustration. It can produce a de facto determination that regional collaboration is too complicated and therefore unworkable.

Yet, the seemingly infinite number of combinations is also a tremendous advantage since it means that there is great potential to devise highly customized approaches that are finely tuned to local circumstances. We believe that the focus on "benefits and issues" as presented and discussed in this chapter, coupled with an understanding of the six specific types of collaboration, offers a much needed path through the forest of possibilities. In any navigation, it is always helpful to have a reference point as a place to begin. By starting with consideration of the mutual benefits from collaboration, potential partners can quickly see how "win/win" possibilities are the best way to start a conversation and keep it going. Turning such conversations next to the issues of (1) how the perceived benefits are to be assured, (2) how costs are to be shared, (3) how joint decisions are to be made, and (4) how other concerns are to be addressed and managed, the dialogue helps determine whether a potential collaboration can or cannot be realized.

In the end, the acid test is whether a written agreement can be reached that explains how win-win benefits can be assured, and potential issues and concerns managed. It is this funneling down process that helps clear a path through the seemingly endless range of possibilities. Working backwards from that endpoint, we observe that there are only six generic categories of collaborative agreements that can result. This certainly thins the forest and, we believe, helps to simplify and clarify the task of evaluating the possibilities. In the next chapter, we use this framework to structure our decision support tool. The decision support tool uses this discrete set of categories as boundaries for consideration, and guides the process of thinking through the benefits and issues associated with each one.

CHAPTER 6
A DECISION SUPPORT TOOL FOR EVALUATING REGIONAL SOLUTIONS

BACKGROUND AND APPROACH TO DEVELOPING A DECISION TOOL

As discussed in chapter 5, the myriad of options for regional collaboration between water systems can be distilled down into six generic types of arrangements that could ultimately fall into place between partnering entities. These six variations are a very practical grouping because they are defined in terms of the type of mutual agreement that has to be reached through negotiations between the partners in order to realize a collaborative solution. The test of whether an agreement can be achieved is a good means of proving the merits of a proposed collaboration. It is also of great practical use to be able to focus discussions within a smaller range of possibilities that can be more readily analyzed and evaluated.

In addition to the six categories of collaborative agreements, there is an additional dimension of variation that must be considered. It is also necessary to decide whether the collaborative strategy is to be applied to the entire operation or only a part of it. Individual options may be applied to source of supply and treatment, to distribution and storage, to management and administration, or to various combinations of all three functional areas of water system operations. This produces a 3×6 matrix of possibilities, as illustrated in Figure 6.1.

As discussed at length in chapter 5, the process of carefully evaluating prospective win-win benefits of a collaboration while also identifying and evaluating collateral issues involved in a collaboration scheme is the much needed path through the woods of the myriad possible options for regional collaboration. But, in order to structure such a deliberative process, it is first necessary to arrive at a determination as to which cell of the matrix in Figure 1 provides the best context for having such discussions. Finding your way initially to the right cell may actually constitute a first screening phase of discussions. It is important to recognize that the details needed to confirm potential benefits or to fully address issues and concerns cannot be completely specified in the screening phase, but must await more specific deliberations once you have settled your agency within a cell of the matrix.

Figure 6.2 depicts the basic approach we adopted to developing a decision support tool that will help a water system first screen and then evaluate options for regional collaboration.

As Figure 6.2 shows, moving from the status quo (the top left of the figure) to a regional solution is a journey that is comprised of several discrete steps. The decision support tool (which is provided on the enclosed CD, and which operates in Excel) is designed to assist water systems in working their way through this process.

In order for regional collaboration to emerge as an idea and progress to a reality, it is first necessary to break away from the inertia of the status quo. There must be compelling reasons to entertain the prospect of making significant organizational changes. As illustrated in the top portion of Figure 6.2, such motivation is often a product of the fact that the current and future business environment are very different than the environment that existed when the existing water systems in the region were first organized. The potential mismatch between the challenges posed by the historical business environment and today's business environment can produce strong motivations to seek an alternative service delivery or organizational model for providing water service.

Regional Solutions Matrix

Water System Functional Areas	Regional Solution Categories					
	Mutual Aid Agreements	Sharing Arrangements	Water Purchase Agreements	Collaborative Regional Water Resource Development	Contract Services Agreements	Consolidation of Multiple Systems
Planning, Development and Operation of Supply & Treatment Facilities						
Distribution System Operations and Maintenance						
Water System Management and Administration						

Figure 6.1 Matrix of regional collaboration possibilities

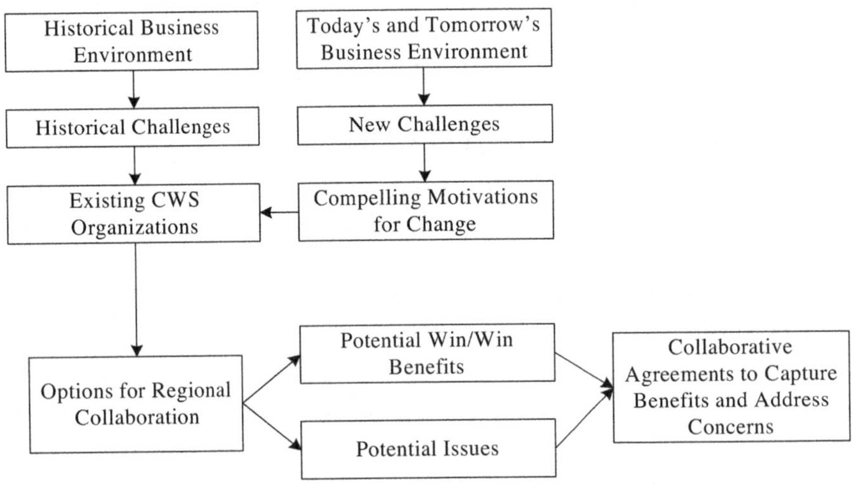

Figure 6.2 Regional solutions roadmap

While there is a wide array of options for collaboration between water systems, the likely starting point consists of options that are most relevant to the challenges that have motivated the search for organizational change. It is useful to be able to focus directly on the most relevant options rather than straying into less desirable arrangements or wandering around in a forest of endless possibilities.

Every option for regional collaboration poses a unique set of opportunities for the participants to derive benefits from the deal. Every option also carries with it a unique set of issues and concerns. These issues can include ways to guarantee that the desired benefits are delivered; control costs in accordance with expectations; and govern the day-to-day functioning of the arrangements to meet performance expectations.

The decision support tool is conceived as a sequence that guides the user through a three-step process:

1. *Examining Motivations:* Identify and rank potential threats and challenges confronting the water system that may exist or may emerge over the near term future.
2. *Choosing Relevant Options:* Map the specific threats and challenges identified as high priorities by the system to relevant types of regional solutions—finding the right cell or cells in the Figure 6.1 options matrix.
3. *Evaluating Benefits and Issues:* Map identified regional solutions to appropriate implementation topics. These topics address the benefits that are needed to make the change worthwhile, as well as the issues and concerns that must be resolved to everyone's satisfaction.

STEP 1—EXAMINING MOTIVATIONS FOR CHANGE

The first step in the decision process involves helping utilities to discover and examine their possible range of motivations for changing existing organizational arrangements. The SWOT (Strengths Weaknesses Opportunities Threats) framework is used extensively in strategic planning and has been used as a tool for conducting this first step in many potential regional collaborations. There is always some double-counting between strengths and weaknesses and opportunities/threats. This is acceptable when facilitating flip chart discussions for stakeholders and getting ideas in the air. It does not work as well, however as a tool framework. Whereas discussions and flip chart outputs can be actively facilitated, the semantics and definitions must be clear and discrete within a self-assessment software tool.

Moreover, strengths and weaknesses do not directly produce an articulation of the motivations for change. In the end, we decided on a Step 1 tool that promotes the desired articulation of a system's motivations in terms of "threats and challenges." Opportunities are not directly addressed at this stage. They are presented once utilities have each identified their motivations and begun the next steps of recognizing the potential for win/win benefits.

The decision tool contains a comprehensive list of threats and challenges covering all of the major functional areas of water system operation:

- Source of Supply
- Treatment
- Storage and Distribution

- Management and Operations
- General Administrative and Financial Management

In conceiving the questions needed to operationalize this self-assessment step in the decision tool, considerable thought was given to the dilemma that there are often two broadly different categories of systems that might be able to see benefit from regional solutions—"unhealthy" systems that are presently facing trouble or difficulty; and "healthy" systems that do not regard themselves as being in any present difficulty, but may be facing future challenges or seeking broader opportunities. In order to accommodate both types of systems within the same tool, the self assessment step is sub-divided into two steps:

- A retrospective assessment of existing problems and challenges; and
- A prospective assessment of potential threats and challenges.

The list of specific problems and challenges included in the retrospective assessment is presented in Table 6.1. It contains items indicative of various types of deficiencies in system performance across the major functional areas.

The items in Table 6.1 are the types of elements that might typically be identified in unhealthy systems in the course of small water system "capacity assessment" reviews now required under the SDWA. Examples include:

- Source of Supply—evidence of supply limitations or possible shortages
- Treatment—evidence of treatment failure or increasing struggle to keep up with SDWA regulations
- Storage and Distribution—evidence of performance failures (e.g., low pressure, taste and odor, chlorine residual or bacterial violations) or increasing struggles with SDWA requirements and aging infrastructure
- Management and Operations—evidence of inadequate resources to do the job (e.g., labor, equipment, management systems)
- General Administrative and Financial Management—evidences of inadequate administrative systems and weak or inadequate financial capacity

The list of prospective specific threats and challenges is presented in Table 6.2. It is drawn in large part from a sense of the broader underlying mega-trends that are sweeping the industry in each functional area. These can have an effect on both healthy and unhealthy systems:

- Source of Supply—source protection; tragedy of the commons; dwindling supply options; reliability of supply strategies; etc.
- Treatment—changing SDWA regulations; aging infrastructure; increased complexity of operations; uncertain capacity needs; etc.
- Storage and Distribution—new SDWA regs measured at the tap; increased complexity of operations; aging infrastructure; etc.
- Management and Operations—increased complexity of operations; human resource needs; reliability; security; OSHA; etc.
- General Administrative and Financial Management—modernizing all basic support service; raising rates as needed

Table 6.1
List of self-assessment criteria for retrospective evaluation

Source of Supply

Water use restrictions have been imposed recently / repeatedly.

Surface water source flows have dropped near or below levels needed to meet our peak demand.

Groundwater sources are being continually drawn down.

Purchased water sources have been interrupted during drought events.

Outages have resulted from failure of key facilities: intakes, tanks, pumps, etc.

Groundwater sources are increasingly vulnerable to sanitary or chemical contamination.

The worst drought of record occurred recently.

Treatment

Boil water advisories have been invoked recently / repeatedly.

Public notification due to a variance or violation of SDWA standards has been required recently.

Violation of SDWA M&R requirements has happened recently / repeatedly.

Violation of MCLs or Action Levels has happened recently / repeatedly.

SDWA treatment requirements are exceeding the capabilities of the current plant.

It is difficult to implement recommendations of latest sanitary survey or CPE.

Plant maintenance activities are difficult to keep up with or have been deferred recently / repeatedly.

Major pieces of treatment equipment have failed recently / repeatedly.

Power outage has caused treatment outage recently / repeatedly.

Storage and Distribution

Low pressure problems have occurred recently / repeatedly.

It is difficult to obtain / sustain pressures needed for fire protection ratings.

Taste or odor complaints have occurred recently / repeatedly.

Power outage has caused water outages recently / repeatedly.

Small mains have broken recently / repeatedly.

System maintenance activities are difficult to keep up with or have been deferred recently / repeatedly.

Major storage and pumping facilities are non-redundant.

Main extensions are constrained by capacity problems.

Management and Operations

Key positions have been vacant or subject to high turnover.

Training is not up to date or has been deferred recently / repeatedly.

Violations of SDWA M&R requirements have occurred recently / repeatedly

Responsiveness and capability to cope with droughts or emergencies has been poor.

Staff is easily overwhelmed when everything happens at once.

Cross connections have occurred recently / repeatedly.

Backflow episodes have occurred recently / repeatedly.

OSHA violations have occurred recently / repeatedly.

Worker's compensation claims are common.

General Administrative and Financial Management

Number or severity of customer complaints is increasing.

Billing errors occur repeatedly.

The level of uncollectibles is increasing.

Contributions to reserve funds have recently / repeatedly been missed.

Debt service payments have recently / repeatedly been missed.

Applications for new loans or credit have recently / repeatedly been turned down.

Table 6.2
List of self-assessment criteria for prospective evaluation

Source of Supply

Land use & development in watershed / wellhead areas may affect recharge rates & sustainable yields.

Additional development of our existing water sources may produce low flow / drawdown impacts.

Surface or groundwater withdrawals by other users may reduce safe yields of our sources.

Long-term reliability / yield of current purchased water supplies may be in doubt.

Current water supply portfolio may be vulnerable to shortage.

Current water supply portfolio is not diversified enough.

Limited or no additional raw water sources exist within our jurisdiction or under our control.

Cost and availability may limit access to purchased water sources in the future.

Cost, geography and acceptance may limit availability of reclaimed wastewater to meet some demands.

Accuracy of demand and conservation forecasts is questionable, leaving capacity estimates uncertain.

Effectiveness of drought management / emergency curtailment is questionable.

Major infrastructure will require rehab or replacement.

Treatment

Land use & development in watershed / wellhead areas may pose new threats to water quality.

Intensity of agriculture in watershed / wellhead areas may pose new threats to water quality.

Our purchased water supplies may be subject to new water quality threats that may increase our cost.

New SDWA regulations may impose an expense to upgrade our treatment plants or those of suppliers.

Existing treatment plant facilities or those of purchased water suppliers may need rehab or replace.

New SDWA regulations may require more extensive and sophisticated water quality monitoring and lab work.

New SDWA regulations may require more extensive and sophisticated recordkeeping and reporting.

New, sophisticated treatment facilities may be needed to meet new SDWA regulations.

Cost of treatment may increase due to the complexity of using multiple source waters.

Costs of treatment may increase due to new requirements for residuals disposal.

Treatment costs may increase due to uncertainty in estimating demand and capacity requirements.

Storage and Distribution

SDWA requirements may require more extensive finished water quality monitoring and lab work.

SDWA regulations may require extensive and sophisticated recordkeeping and reporting.

SDWA regulations may require significant operational changes to meet targets measured at the tap.

The age & condition of small mains may cause water quality, pressure and repair problems.

Major infrastructure may require rehab or replacement.

Control systems should be expanded or modernized.

Population changes may affect fire fighting needs and peak capacity/pressure requirements.

Backflow prevention and cross-connection control efforts may require more effort to remain current.

Policies for extensions and replacements may need to be modified to manage changes in the system.

(continued)

Table 6.2 (Continued)

Management and Operations	
	Meeting paperwork and M&R requirements may be difficult.
	SDWA regulation compliance may become a strategic issue affecting major capital decisions.
	Staffing levels and staff training needs may escalate.
	Changes in workforce demographics may pose a significant challenge for succession planning & knowledge retention.
	Needs for new equipment (trucks, backhoes, etc.) may increase due to system growth and/or equipment age.
	Reliability may be affected in some areas where there is no redundancy for key system components.
	Back-up power supply is incomplete or of questionable reliability under some scenarios.
	Emergency preparedness program requires modernization to keep up with new challenges.
	System security program requires actions to implement recommendations and assure continual vigilance.
	OSHA compliance program and overall safety program may require modernization.
	General maintenance activities may need to be shifted to more proactive and less reactive tasks.
General Administrative and Financial Management	
	Public relations and public information may need greater recognition as a formal activity of management.
	Customer service and billing/collection functions may need to be reviewed and updated.
	Meter testing, maintenance and replacement program may need to be reviewed and updated.
	Budget and capital planning, as well as rate reviews, may need to be reviewed and updated.
	Risk Management (legal reviews and insurance) may need to be reviewed and updated.
	Human Resources policies and practices may need to be reviewed and updated.
	Procurement and expenditure control policies and practices may need to be reviewed and updated.
	We may have trouble keeping operating cash reserves at adequate levels.
	We may have trouble keeping debt loads at manageable levels that are within market and legal limits.

The Step 1 tool that emerged is one that is more appropriate to the small and mid-size systems. Large regional entities (>75,000–100,000 population) that might be extremely important to a regional scheme might not see themselves in some of the threats and challenges from the way they are articulated here. They may have the reciprocal problem in which their surplus capacity (physical or managerial or financial) is just the right solution to a smaller system's deficiency. Their biggest problem might be getting the smaller systems to see the bigger picture and this tool may therefore be of use to them from that perspective. They could use it to screen for potential partners or as a tool to help potential partners build support for collaboration. In a way, one of the biggest threats to the larger systems is to miss opportunities to grow by providing services to neighboring systems.

To complete Step 1, the tool asks the user to assign a priority ranking to each of the conceivable problems, threats, and challenges on the list. This will be accomplished by entering a #1 in the columns to the left of each item, choosing between a priority ranking of high, medium, low, or not applicable. The spreadsheet can then be summed to produce a summary assessment of the top priority items. From this summary, the user will obtain a sense of how the pattern of their priorities falls across the major functional categories: Source of Supply, Treatment, Storage and Distribution, Management and Operations, General and Administrative and Financial Management. These patterns of priority motivations will be used to "map" the user's results into the Options Matrix to complete Step 2 in the decision analysis procedure.

Regional Solutions Matrix

Water System Functional Areas	Regional Solution Categories: Mutual Aid Agreements	Sharing Arrangements	Water Purchase Agreements	Collaborative Regional Water Resource Development	Contract Services Agreements	Consolidation of Multiple Systems
Planning, Development and Operation of Supply & Treatment Facilities	Emergency Interconnections & Collaborations in Emergency Response	Joint Purchasing and Sharing of Specialized Resources	Contract Purchase of Treated Water to Augment or Supplant Own Supplies	Joint Planning, Development and Operation of Supply and Treatment Facilities	Contractor Provision of Water Supply & Treatment O&M Services	Merged Ownership and Operation of Supply and Treatment Facilities
Distribution System Operations and Maintenance	Collaboration in Emergency Response	Joint Purchasing and Sharing of Specialized Resources			Contractor Provision of Distribution System O&M Services	Merged Ownership and Operation of Distribution Systems
Water System Management and Administration		Shared Management and Administration Functions			Contractor Provision of Management and Administration Functions	Merged Management and Administration Functions

Figure 6.3 Regional solutions options matrix

STEP 2—CONSIDERING OPTIONS FOR REGIONAL COLLABORATION

Figure 6.3 presents the Options Matrix worksheet of the regional solutions decision tool. This matrix contains all of the generic types of regional solutions on one axis mapped against the major areas of priority need on the other axis. The user will initiate Step 2 by identifying the specific cells within the options matrix that are most relevant to their priorities.

In general, it is the intent to guide the user to start by looking at options that match their highest priority needs or motivations. So, if the preponderance of priority rankings land in the Source of Supply and Treatment areas of the self assessment worksheet, the user would be directed to look first at the Source of Supply and Treatment row of the Options Matrix. Although it is certainly possible to put together combinations of options across the rows and columns of the matrix (e.g., purchased water + contract operations of storage and distribution), it is the intent to start with the highest priority cells of the Options Matrix and examine them one at a time.

STEP 3—EVALUATING BENEFITS AND ISSUES

Once focused on a cell or set of cells within the Options Matrix, the user can activate Step 3 of the procedure—the evaluation of benefits and issues—by clicking a mouse on the hyperlink title within each cell of the Options Matrix. Clicking on the hyperlink will call up a brief

PowerPoint presentation that contains a summary of the benefits and issues relevant to the option indicated in that cell of the Options Matrix. The PowerPoint slides:

- Describe the option in terms of the simple mechanics of the collaboration involved.
- Describe the intended win/win benefits to be derived by all participants; addressing have vs. have not issues and larger outcomes for the economy and environment within the region as a whole.
- Describe the common forms of this approach to regional collaboration that appear among the examples that exist. This includes a review of common types of organizational relationships, alternative institutional structures, and types of agreements needed to formalize arrangements.
- Provide an overall summary of the issues and concerns that could impede implementation and suggested approaches to resolving them. These include issues of cost control, state and local legal questions, labor force considerations, asset transfer issues, growth vs. no-growth issues, performance guarantees, water rights, limitations of liability, and other topics.
- Provide a summary "suitability checklist" that characterizes the likely circumstances of a utility that might expect to derive the most benefit from each regional solution option.

The PowerPoint slides summarize the essential points needed to provide support in undertaking negotiations with potential partnering water systems.

USING THE DECISION SUPPORT TOOL

The decision support tool described above can be found on the enclosed disk. The computerized version can be opened and run in Excel, and has been made user friendly through the use of a Visual Basic interface built into the spreadsheet tool. For further guidance on how to access and use the tool in computerized form, see the "Read Me" page in the file.

CONCLUSIONS

The myriad types of potential regional approaches to water supply cooperation can make the process of considering a consolidation effort seem unwieldy. Further, the process of considering a possible regional approach needs to originate within the context of understanding what challenges are most pressing to a water utility and its neighbors. The decision support tool described in this chapter provides a framework and easy-to-use process to help guide a water agency (or other interested parties) through the process of identifying and assessing challenges and needs, and then helping to match the utility's context to the subset of regional approaches that may be most suitable for consideration. This provides a useful process of self assessment and winnowing of options, so that efforts can then be focused on a practical range of best-fit alternatives.

In the next chapter, we provide summaries of several case studies of regional approaches that have either blossomed or been declined, as compiled by the research team in conjunction with the project's participating water agencies. These case studies reveal the challenges and needs that have driven several areas around the U.S. to consider regional approaches, describe the consolidation options contemplated and (where applicable) implemented, and identify what factors and considerations helped to either promote and sustain a working collaboration, or contributed key obstacles to the pursuit of a regional effort. These case studies complement the decision support tool developed here, by offering real examples and distilling key lessons learned from efforts to implement regional solutions for water supply services.

CHAPTER 7
CASE STUDIES

This chapter provides summaries of several case studies the research team developed in conjunction with participating utilities and a review of the literature. These case studies provide examples of the motivations for and types of regional approaches that have been attempted and, in most cases, implemented within the U.S.

In this chapter, we provide abbreviated versions of the case studies; the full-length case studies are provided as appendices for the more in-depth and quantitative case studies for Portland, Oregon, and the North Carolina central coastal plain. In this summary, the research team focuses on the key points for each case study, including:

- The motivation for exploring regional approaches in that case
- What regional approaches were explored and adopted (or rejected)
- Key factors contributing to the implementation (or lack thereof) of the regional approach
- Key barriers to be overcome (and how they were or were not surmounted)
- Major lessons learned

Overall, the intent is to distill central themes, user tips, and key lessons learned from the case studies. For example, we note that (1) realized regional approaches have most often been driven by clear lack of adequate quantities of water supply, (2) regional approaches appear more palatable if each party retains ownership of certain key assets, and (3) retaining some measure of local autonomy is a critical issue that frequently determines whether a regional approach will be considered favorably. A summary of the case studies is provided in Table 7.1, and each is provided below, in alphabetical order.

CHAMPLAIN WATER DISTRICT

Background

The Champlain Water District (CWD) began operating in 1973, and is located in South Burlington in northwestern Vermont. The source of water is Shelburne Bay, on the eastern shore of Lake Champlain in Chittenden County. CWD is a medium sized provider serving a population of approximately 68,000 and averaging 10 million gallons per day (mgd) from a single surface water treatment facility. CWD wholesales water to 12 municipal consecutive water systems in nine communities, and all of the systems are physically interconnected. In addition to water provision, the CWD provides various levels of retail support services such as meter reading, billing services, and maintenance support to five of the 12 municipal water systems. As the wholesale supplier, CWD also conducts all SDWA regulatory compliance involving source protection/treatment requirements, as well as serving as the responsible party for all SDWA compliance sampling and reporting requirements related to distribution systems for the twelve served municipal water systems. CWD operates as a municipally chartered "Consolidated Water District" under Vermont enabling legislation.

Table 7.1
Summary of case studies

Case study region and key entities	Motivation for exploring and/or implementing regional approach	Type of regional approach used or attempted (and outcome)	Key factors contributing to success	Key obstacles and factors limiting success
Champlain Water District				
• Serves approximately 68,000 in South Burlington in northwestern Vermont with an average of 10 mgd. • The source of water is Shelburne Bay, on the eastern shore of Lake Champlain in Chittenden County. • One surface water treatment facility.	• Several towns in Chittenden County had similar water problems, either with source quantity or with water quality.	• CWD currently wholesales water to 12 municipal consecutive water systems. All systems are physically interconnected. CWD also provides retail support services such as meter reading, billing services, and maintenance support to 5 of the systems. Via an executed agreement, CWD is also the SDWA compliance entity for all of its 12 served municipal water systems.	• Personal vision and drive, and government support are key factors to the success of the CWD. • The State of Vermont had enabling legislation for the creation of consolidated water districts, and helped CWD obtain a $3 million federal grant. • Proactive approach to regulatory challenges and compliance for all served municipal systems performed by CWD.	• Authorities of county government are very limited, whereas authorities of local governments are very strong. Sometimes this creates challenges for proposing and implementing regional solutions. • Developing and implementing a watershed management program for source water protection in the Shelburne Bay watershed may offer new challenges because only 3 of the 8 municipalities involved are members of CWD.

(continued)

Table 7.1 (Continued)

Case study region and key entities	Motivation for exploring and/or implementing regional approach	Type of regional approach used or attempted (and outcome)	Key factors contributing to success	Key obstacles and factors limiting success
Lehigh County, Pa.				
• Lehigh County Authority (LCA), Allentown, Pa. • LCA serves 16,000 water customers, 1,500 wastewater customers, and provides wastewater transportation service for 10 municipalities. • LCA borders the City of Allentown (with its own water and wastewater systems), as well as numerous smaller public and private water and wastewater providers.	• Seeks to increase efficiency of operations; maximize use of existing personnel, expertise, and facilities. • Dramatic population shifts resulting in new development in outlying areas, with underutilization of utility facilities in older, more urbanized areas. • Management succession issues in LCA and many neighboring systems, as managers and other key employees approach retirement.	• Acquisition of smaller, neighboring systems • Success with several privately owned systems (primarily developer-built systems). • Little success with publicly owned systems. • Consolidation with large public system(s) • Promising talks stalled due to change in political control. • Process may be reinvigorated due to interest and funding from regional economic development partnership.	• Longevity and solid reputation in community. • Strong management team with in-house legal and engineering expertise (reduces cost of pursuing acquisitions or other forms of consolidation).	• Concerns with loss of local control. • Use of water systems as de facto method for land-use control. • Achievable cost savings lower than might appear, due to sharing of services between water and other municipal departments. • Small systems' lack of appreciation of risks of continued ownership and operation of system (e.g., lack of knowledge of future regulatory requirements). • Decentralized government (lack of regional planning, county government has little power over local governments). • Many publicly owned systems subsidize general government expenses from water/ wastewater funds.

(continued)

Table 7.1 (Continued)

Case study region and key entities	Motivation for exploring and/or implementing regional approach	Type of regional approach used or attempted (and outcome)	Key factors contributing to success	Key obstacles and factors limiting success
Massachusetts Water Resources Authority (MWRA)				
• Provides wholesale water and wastewater services to 2.5 million people in 61 communities in the Boston metro region of eastern Massachusetts.	• MWRA predecessor (Metropolitan District Commission) initially established due to insufficient quantity of source water as the City of Boston and neighboring communities grew in population and economic activity. • MWRA itself established as a result of federal court case concerning water quality problems and the pollution of Boston Harbor and need for autonomous agency to address supply, quality and environmental requirements.	• MWRA and predecessor organizations established to provide wholesale water and sewer services to member communities. • Predecessor could not sustain provision of regional water and sewer services, while meeting growing demands and environmental regulations. • MWRA constructed new water and wastewater treatment plants; cleaned up Boston Harbor; also reduced water demand/use significantly (25%) to within safe yield of system and implemented long term planning and capital improvement program.	• Management and financial autonomy allowed MWRA to raise required revenues to address long-neglected maintenance and capital improvement needs. • Court order provided political impetus for establishment of new agency; cooperative approach by wholesale agency didn't threaten local municipal systems. • Participation and oversight by member communities and other stakeholders created accountability and legitimacy. • Integrated resource planning (supply and demand; water and wastewater) enhanced overall effectiveness.	• Predecessor agency lacked managerial and financial autonomy; funding was subject to political decisions of Legislature. • Huge capital needs of long-neglected system required politically difficult, large rate increases.

(continued)

Table 7.1 (Continued)

Case study region and key entities	Motivation for exploring and/or implementing regional approach	Type of regional approach used or attempted (and outcome)	Key factors contributing to success	Key obstacles and factors limiting success
Metropolitan Water District of Southern California (MWD)				
• MWD is the largest water supplier in the U.S. It provides wholesale water to 18 million people in 26 cities and water districts in parts of Los Angeles, Orange, San Diego, Riverside, San Bernardino, and Ventura counties.	• MWD was formed for the purpose of developing and distributing water from the Colorado River in order to provide a sufficient and reliable water supply to the growing population of Southern California, while protecting the independence of smaller municipalities from the dominating political control of Los Angeles.	• MWD was established to provide supplemental water supplies to its member districts, initially from the Colorado River. • MWD provides its member communities with additional water rights and the capital with which to develop these additional supplies.	• Through federal loans and aids, the MWD was able to successfully market the bonds to raise $220 million for the Colorado River Aqueduct. • MWD is an independent local unit of government. It is governed by a board of directors that consists of 37 members representing all the member agencies. • Member agencies remain as retailers of the water, retaining their assets and customers.	• Lack of coordination among MWD and local water agencies threatened to worsen the water shortage, as the local agencies had independent water supply strategies that were not always compatible with MWD's regional plan.

(continued)

Table 7.1 (Continued)

Case study region and key entities	Motivation for exploring and/or implementing regional approach	Type of regional approach used or attempted (and outcome)	Key factors contributing to success	Key obstacles and factors limiting success
North Carolina Central Coastal Plain				
• 15-county rural region, relying primarily on local groundwater to serve a large number of small communities. • See extended case study in Appendix B.	• Insufficient quantity of source water. • Unsustainable over-extraction of regional groundwater revealed alarm about potential future loss of aquifer usability. The State of North Carolina established 3-step pumping phase-down plan that will reduce withdrawals by 30% to 75% for every local community. • Replacing groundwater with surface water is very expensive for most communities due to treatment needs and conveyance distances.	• New regional approaches being developed for multiple surface water-based regional provider. These should successfully deliver treated surface water to many of the impacted small CWSs, and provide sizable cost savings. • Tradable groundwater pumping permits also being explored, as an innovative mechanism to offer additional cost savings across the region (when implemented in concert with regional surface water systems).	• Willingness of local communities, including those with existing surface water supplies, to pursue cooperative ventures to provide more surface water to neighboring communities. • Willingness of state and county to provide technical resources and financial advice to local communities. • Support for, and availability of research that reveals the overall size of the benefits (cost savings) that can be realized by using regional approaches. • State regulatory agency mandated significant reductions in groundwater pumping.	• Initial difficulties convincing local communities that there was a groundwater problem that needed to be solved. • Initial efforts by state to allocate pumping restrictions were resisted due to lack of transparency about the rules and the situation. Later efforts rectified this problem. • Many small communities impacted, with limited resources.

(continued)

Table 7.1 (Continued)

Case study region and key entities	Motivation for exploring and/or implementing regional approach	Type of regional approach used or attempted (and outcome)	Key factors contributing to success	Key obstacles and factors limiting success
Portland, Ore.				
• City of Portland, Ore. (CoP), Portland Water Bureau (PWB), and 13 of 21 metro area water providers serving suburban communities. • PWB serves 786,700 customers, including 482,500 Portland residents and 300,000 metro area residents through wholesale contracts to 21 area water providers. • See extended case study in Appendix A.	• Opportunity for shared ownership of/decision making around main water supply source (the Bull Run watershed). • More efficient planning/coordination and concomitant cost savings. • Diversify/expand sources to mitigate possible shortage risk.	• PWB and 13 water providers entered into an agreement to explore creation of a regional water agency with responsibility for protecting the region's water supplies, maintaining distribution systems, and planning for future water needs. • After 1.5 years of research and public feedback, CoP withdrew support for the initiative. Negotiations commenced to replace soon-to-expire long term contracts with water providers. Large amount of data and improved understanding of area water needs generated by regionalization initiative will inform these negotiations.	• Regional approach was not altered; prior water purchasing contracts being renewed instead. • While effort did not result in a change in the regional approach, the process of exploring options provided useful knowledge and will help in future efforts.	• Buy-in cost to compensate CoP for existing source/ transmission infrastructure considered too high by regional partners. • CoP hesitant to relinquish PWB and its revenue-generating capacity to regional agency; key CoP financial stakeholders not involved early enough in plan. • Portland residents' concern over the quality of alternative water sources being considered for inclusion in the agency ultimately impeded consideration of a fully "regional" agency with regard to area water supplies.

(continued)

Table 7.1 (Continued)

Case study region and key entities	Motivation for exploring and/or implementing regional approach	Type of regional approach used or attempted (and outcome)	Key factors contributing to success	Key obstacles and factors limiting success
Central Puget Sound, Wash.				
• Central Puget Sound Water Suppliers' Forum, Wash. • Serving 3 city water utilities (Tacoma, Seattle, and Everett), 3 county water utilities (King, Pierce, and Snohomish), and 8 other associations of water utilities, in total serving 2.5 million people.	• Uncertainty in implications of requirements of the Endangered Species Act (ESA) and the Growth Management Act (GMA), coupled with growing municipal needs to supply water to growing populations.	• New regional entity (the Forum) developed to coordinate water supply planning, but not govern. • Most uncertainties with the GMA have been addressed by the Forum. However, some small systems may be unsure if they have sufficient long term supplies. • ESA uncertainties (e.g., who is responsible for providing water, how much is to be provided) remain unresolved. The Forum is addressing the issue, but not taking a pro-active role because of a sense that water utilities should not lead this effort and should let the federal and/or state government and tribes lead. However, many member utilities are comforted by having an organized regional body to collectively engage the issue.	• Local CWSs retain all asset ownership and customer interface. • Common external challenges (projected water supply needs and ESA and GMA requirements). • Personal leadership and drive within the utility industry.	• Autonomy is a double edged sword—allowing members to retain autonomy facilitates participation but creates challenges (it can be difficult to get all participants to agree). • Since the Forum is comprised of different entities with different political issues, sometimes it is difficult to align them. • Maintaining participant interest in regional approaches absent a threat or major project can be a challenge.

(continued)

Table 7.1 (Continued)

Case study region and key entities	Motivation for exploring and/or implementing regional approach	Type of regional approach used or attempted (and outcome)	Key factors contributing to success	Key obstacles and factors limiting success
Tampa Bay Water, Fla.				
• Provides 172 million gallons of water a day to over 2 million people in the counties of Hillsborough, Pasco, and Pinellas; and the cities of New Port Richey, St. Petersburg, and Tampa.	• Unsustainable/over extraction of regional groundwater caused severe environmental and property damage. • Population growth caused the need for additional water supplies.	• New regional authority developed, successfully supplies member agencies with wholesale treated surface water (including desalted sea water) along with sustainable levels of groundwater. • Regional groundwater extraction reduced to acceptable levels. • Ended the Tampa Bay water wars (which cost the region millions of dollars in litigation). • Water supplies will need to increase by 8 to 12 mgd to meet future demand. • Approach is generally succeeding to provide a reliable regional supply and preserve groundwater resources.	• Local CWSs retain most asset ownership and customer interface. • State regulatory agency mandated significant reductions in groundwater pumping. • Significant infusion of state (and some additional Federal) funding. • All members pay the same price for water and share the costs of developing new water sources. • Pooled taxes enabled the development of more costly alternative water supply projects.	• Lack of county-level authority to manage or guide localities. • New surface water supplies very expensive relative to old sources.

Motivation for Considering Regional Solutions

In the mid-1960s, several towns in Chittenden County had similar water problems, either with source quantity or with water quality. Burlington (then population approximately 27,000) was supplying water to a few towns such as South Burlington (then population approximately 10,000). At that time, South Burlington desired more control over water rates and water quality. Also, several towns in Chittenden County surrounding South Burlington had unfiltered surface water supplies and unreliable groundwater supplies as well, but never discussed water issues with each other (AWWA 2004).

Steps Toward Consolidation

Often times, personal vision and drive spearhead movements towards regionalization. That was the case with the CWD. In the mid-1960s, the Board Chair of the South Burlington Water Commissioners, Peter Jacob, began approaching other municipal water boards in the county about discussing their mutual future growth and water issues. In the spring of 1966, water commissioners from Essex Junction, Williston, and Charlotte met with the South Burlington Water Commissioners to discuss forming a consolidated water supply district. By an overwhelming majority vote in September of that year, those three towns became members of what was later named the Champlain Water District.

Another significant driver toward regionalization is government support. The CWD had this component on their side as well. The State of Vermont had enabling legislation for the creation of consolidated water districts, which they, as well as the federal government, much preferred over creating, funding, and working with several individual municipal water treatment facilities. They also received considerable help from then Vermont Senator George Aiken, who was instrumental in getting a $3 million grant approved for CWD from the U.S. Department of Housing and Urban Development. This enabled construction to begin on a treatment plant, intake in Lake Champlain, pipelines, and storage tanks in 1971 (AWWA 2004).

Structure of the CWD

The Board of Water Commissioners of the CWD has eight members, with one member publicly elected from each town for a three-year term. One alternate from each member town is also appointed by each town to serve if the elected Commissioner is absent. The Board is not weighted by population or water use; each town, regardless of size, has one vote. In addition, all residents of the town vote for the water commissioner, not just those residing in areas served by CWD water, since the town is the member and owner of the overall District.

The towns wanted to maintain their existing physical assets, so the CWD was created with flexibility. The towns maintain ownership of their distribution and storage infrastructure while the District owns and maintains the transmission supply infrastructure. Technical functions were combined across the system, but local control was still available for water systems that preferred to manage tasks such as billing, operation and management. Therefore, the CWD maintains a network of water transmission mains, storage tanks, and booster pump stations strategically located within the CWD Chittenden County service area. CWD Treatment and Transmission Specialists operate the water treatment facility and the countywide transmission and storage network on a 24/7/365 basis. There are presently 21 major water storage tanks and six booster pump stations that CWD operates and controls. Four water storage tanks and two of the six pump stations are still owned and maintained by two of the twelve served water systems, although deeding storage to CWD for preventive maintenance is allowed. Once storage is deeded to CWD, required maintenance and recoating is performed as part of the overall wholesale water rate, while requirements for new expanded storage volumes are funded by the served water system.

Although no utilities to date have chosen to leave the water district, the CWD has an exit strategy available for those utilities that may opt to depart. In the enabling legislation and charter, it is stated that if a utility decides to leave the water district to supply their own water needs with a well for example, there is a "buy-out charge" for outstanding debt based on percentage of water use. If one's utility uses 10% of CWD's total water supply, then the buy-out charge would be equal to 10% of CWD's total outstanding debt.

Factors Contributing to the Lasting Success of the Consolidation

For over 30 years, CWD has been a successful regional water provider with an excellent service record. It has been recognized by the American Water Works Association (AWWA) and USEPA for the Excellence in Water Treatment Award under the Partnership for Safe Water. In 1999, the CWD became the first utility in the Nation to earn USEPA's Phase IV Excellence in Treatment Award. At the 2004 AWWA Annual Conference, the CWD received a five-year Phase IV Excellence in Treatment anniversary award.

Seven key aspects of the consolidation effort have been identified as contributing to CWD's enduring success:

A. Ensuring all community participants gained from consolidation (without losing anything)
   a. Federal and State government support (legislative and financial)
   b. Flexibility in membership services
B. Allocating one vote to each member town
C. Focusing on being technical in protecting public health, and not political
D. Having personal drive and commitment from the Board and employees
E. Having strong staff training
F. Implementing procedural and process improvements well in advance of regulatory deadlines and maintaining all SDWA compliance requirements for all 12 municipal served water systems
G. Legally establishing a Capital Reserve Fund fed from annual budgetary surpluses, connection fees, and an annual budget expenditure allocation referred to as "Encumbered Cash Reserves"

Each of these is discussed in turn, below.

A. Consolidating Gains Without Losses

Each utility was eager to join the CWD because they were offered access to new government finances and ways of solving problems potentially unavailable to them individually. However, each water system utility did not want to give up ownership of certain physical assets in exchange for this access. The arrangement allowed towns to retain ownership of their assets.

a. Government Support: Because consolidation in this case meant access to state and federal funds, all utilities clearly were financially better off by joining together. The federal government offered CWD a large amount of capital, while the state offered smaller matching grants. Also, for the consolidation to be effective, it must be recognized by the government as the voice for all of the representing communities. The State of Vermont was eager to give the CWD this voice, as having dialogue with one entity is easier than with many, and it is potentially less costly for the government to offer financial support for one centralized treatment plant instead of several.

b. Flexibility in Membership Services: One key aspect to CWD's creation and success is the flexibility built into the system. Prior to consolidation, water systems expressed concern about losing their physical assets. Therefore, CWD was organized so that members may retain local control while participating in a system where technical functions are combined. The District assumed the role of a water supplier and transmission/storage operator while giving flexibility to local entities for management and ownership of existing water distribution and storage infrastructure.

B. Allocating One Vote to Each Town

Each town has one publicly elected commissioner on the board, with one vote. A system that has 500 connections has the same voting power as a system that serves 5,600 connections. This has proven to work well for the CWD. Other options considered include weighting the board by population or by amount of water use, but giving one voice to each town has proven successful. This issue has arisen at times over the past 30 years, and each time the CWD Board has chosen to remain with their current system, and credit this system for helping the district be so proactive and effective.

C. Focusing on Technical Controls Instead of Political

The CWD board has made efforts to remain a technical group. Their efforts are directed towards effective water supply and management that is protective of public health, and having this common goal among all the members helps keep things running smoothly. The CWD board has made great efforts to treat each issue as a regional district issue instead of an individual local (town) issue. The main focus of the District is what is best for the District overall and not individual member towns. It is felt that whatever is best for the District will also benefit the towns as well. It does not want to be a governing board that changes when the local government changes. The District has a municipal charter that gives them one task and only one task—to provide the safest drinking water possible at an economic rate. This is key since most conventional municipal water departments are competing with other municipal departments for funding and overall recognition.

D. Personal Drive

The current CWD Board Chairman is the same person who initiated consolidation nearly 40 years ago. When the District celebrated its 25th anniversary in 1993, it dedicated the water treatment plant to Peter L. Jacob, in recognition of his 24 years as Chairman of the Board. "It was very appropriate and meaningful to dedicate the district facility to the person who worked tirelessly as a leader to form and guide the development of the regional water supply" (AWWA 2004).

E. Highly Trained Staff

The CWD Board and staff have significant experience. The average board member has been on the board for over 16 years, and the average employee has been with CWD for 12 years. CWD management is a strong believer in staff training with AWWA and its New England section (New England Water Works Association, NEWWA) training programs and those offered by the state and other drinking water organizations. The CWD treatment and transmission specialists continually set water quality goals well beyond the regulatory requirements. "There's a friendly spirit of competition among them, and none of them want these strict self-imposed goals to be exceeded on their shift. They have a lot of pride in what they do, the Board sets high standards for its employees and insists they be well trained and well paid for their efforts" (AWWA 2004).

F. Proactive Approach to Meeting Regulatory Guidelines

Management, Board, and staff habitually plan and implement cutting edge, state of the art solutions to protect public health well ahead of regulatory schedules. CWD also performs all SDWA compliance and reporting for its 12 municipal served water systems.

G. Reinvestment of Revenues Solely for District Future Needs

CWD is not a city or town "revenue department" within the overall typical municipal structure. Since CWD has a regional municipal charter, its revenues are not siphoned off by the overall municipal entity, as payment in lieu of taxes, to hold down the overall city property tax rate. CWD feeds its legally established Capital Reserve Fund with annual budgetary surpluses, connection fees, and an annual budget expense allocation referred to as Encumbered Cash Reserves (2%–5% per year) to reinvest in capital assets with an annual dollar goal equal to annual depreciation. With this approach it is hoped that current customers pay the full cost for the water they receive. The present uniform wholesale water rate is $1.24/1,000 gallons.

Obstacles

The CWD has not had many significant obstacles in forming and in operating, but some of the issues it has faced are briefly described below:

- From time to time, questions arise from state and national regulatory agencies regarding regulation of CWD's specific municipal wholesale customers. As consecutive water systems, regulatory issues fall under the umbrella of CWD's overall compliance structure. However, all agencies realize it is easier to work with one supplier rather than twelve water departments, and recognize that CWD is extremely effective at implementing and assuring compliance under the SDWA.
- Long range planning and overall operation and maintenance practices (e.g., unidirectional flushing, valve maintenance) are not standard for the entire CWD system. Rather, they vary by the size of the served municipal system and are based upon the individual water systems' resources and individual organizational structure (e.g., fire district, water department, public works department).

- Lack of recognition outside of the served communities due to CWD's regional duties and approach in an area of the country where local government is heavily relied upon (and the authority of county government is very limited).
- Some constituents did not think the CWD should have assisted and served new water systems members in the early 1990s. However, the board felt a sense of duty to provide new members in need with good water because they had the capabilities to do so (AWWA 2004). Continuing to extend CWD services beyond it's current membership will be challenging due to economic and geographic restrictions and local/state government planning decisions.
- Developing and implementing a watershed management program for source water protection involving the Shelburne Bay watershed may offer new challenges because of the eight municipalities involved within the watershed area, since only three are members of CWD.

Conclusions

Similar water problems, either with source quantity or with water quality, brought municipalities together and led to the formation of the CWD. For over 30 years since its existence, CWD has been a successful regional water provider with an excellent service record. The CWD is a useful example of a regionalization success story. Elements contributing to its success include:

- Common problems shared by several neighboring utilities
- Visionary and effective leadership
- State and federal support (including financial support)
- Supplying wholesale water only (with towns keeping their own assets)
- Having the towns own the water district
- Towns being able to provide some local water-related services themselves or opting to have CWD provide them (e.g., billing, O&M)
- Keeping a technical focus and not a political one
- Maintaining a proactive approach to meeting regulatory challenges, as well as performing all SDWA reporting and compliance for the 12 served municipal water systems
- Legally establishing a Capital Reserve Fund fed from annual budgetary surpluses, connection fees, and an annual budget expenditure allocation named Encumbered Cash Reserves to continually reinvest in required capital assets

LEHIGH COUNTY, PENNSYLVANIA

Background

The Lehigh County Authority (LCA) was formed in 1966 and provides water and wastewater service in the greater Allentown area in eastern Pennsylvania. Originally, LCA was designed to serve areas in Lehigh County, outside of the City of Allentown, that were not already receiving public water service and centralized wastewater service (EMA, Inc. 2004).

In its nearly 40-year existence, LCA has continued to expand. At year-end 2005, it provided water service to approximately 16,000 customers with an average demand of 5.4 mgd.

LCA also provides a variety of wastewater services, including wastewater transportation (a wholesale service) for 10 municipalities that run their own collector systems, wastewater collection, and transportation for approximately 600 retail customers, and wastewater treatment for an additional 600 retail customers (LCA 2006).

LCA's water service is provided primarily from groundwater supplies. LCA has one major service area for water service serving more than 14,000 customers, as well as eight smaller areas each of which serves between 5 and 1,100 customers. Each area has its own sources of supply and the areas generally are not interconnected with other LCA service areas (though some have interconnections with other public water suppliers) (LCA 2006).

Motivation for Considering Regional Solutions

With the exception of the City of Allentown and LCA, the provision of water and wastewater service in Lehigh County is primarily provided by small or moderately sized boroughs and townships, as well as a few privately owned water systems that tend to be very small (primarily developer-built systems). LCA, county planning officials, and several of the municipalities in the county recognize that small, separately owned and operated systems cannot capture the economies of scope and scale that a regionalized approach would enjoy (Stratus Consulting 2000).

In addition, many of the smaller systems are facing increased cost pressures as a result of new (and prospective) water testing and treatment requirements under the SDWA. Adding to this pressure are two significant factors: (1) the population centers of the county are changing dramatically, and (2) many of the managers of the water and wastewater providers in the county are nearing retirement age.

The first factor—population shifts—is resulting in tremendous growth in outlying areas of the county, with concomitant demand for water and wastewater services. At the same time, the older boroughs and cities in the county (e.g., Allentown) are seeing essentially no growth in population, often coupled with a declining industrial and commercial base, leaving many of them with unused water and wastewater capacity. For example, Lehigh County's cities saw a population increase of just 1,400 people (0.7%) from 1990 to 2000, while the outlying townships grew by more than 34,000 people (17.4%) during the same period, with some townships growing by 50% or more during this period. Other important demographic changes that can affect the demand for water and wastewater services include a decrease in the average number of people per household, an increase in poverty, and a dramatic increase in the portion of the population that is elderly (Brookings Institution 2003).

From a water supply perspective, this demographic trend means that growing suburban and quasi rural areas are investing to expand their supply capacity to meet growing demands, while at the same time older, more urbanized portions of the county are struggling with the cost of excess capacity. This suggests that a more region-wide perspective might reveal opportunities for greater efficiencies and overall cost savings.

The second factor—management succession—also is a significant factor in encouraging the county's water and wastewater suppliers to look for regional solutions. Within the next five to ten years, nearly every significant water and wastewater system in the County is likely to have at least one of its key people retire. Rather than attempting to find replacements for all of these highly skilled people, it may prove to be more cost effective to hire a regional manager, or a few sub-regional managers, to manage the County's diverse water and wastewater systems (Stratus Consulting 2000).

Steps Toward Consolidation

LCA's water service area has grown steadily through a combination of increased development within its existing service area and the acquisition of predominantly small water systems. From 1993 through 2005, LCA more than doubled the number of water customer connections that it serves (LCA 2004, 2006).

That growth trend accelerated after 2000 when LCA commissioned a strategic planning study to help it determine whether and how it should continue to grow (Stratus Consulting 2000). Upon reviewing that study, LCA's Board of Directors decided to implement a more aggressive strategy of encouraging acquisitions and other relationships with neighboring water and wastewater systems. As a result, LCA has acquired several small water and wastewater systems, leading to an increase in its water customer base by more than 58% from 2000 to 2005 (LCA 2006).

LCA also has been actively involved in regional planning efforts with other water and wastewater utilities. For example, LCA participates with the City of Allentown and other municipalities in jointly planning for an increase in the city's wastewater treatment capacity.

While a proactive, regional perspective on the part of LCA has helped promote some regional approaches, there are long-standing state-level institutional constraints that inhibit broader interest and use of regional solutions. A key impediment is the manner in which Pennsylvania law assigns many authorities to local government entities (cities, townships and boroughs), and in contrast provides very little authority to county-level entities. In this institutional setting, there are limited county-level resources or authorities that might otherwise enable planners or elected officials to provide greater region-wide perspective, or to assert more leverage in promoting regional approaches to water supply.

In addition, LCA and Allentown have studied the possibility of consolidating their utility operations. Political discord in Allentown caused the discussions to stop at least temporarily. There are indications, however, that discussions may resume, as a combination of changes in leadership within Allentown and a new initiative from a regional economic development partnership to study the regionalization of utility services.

Structure

LCA was formed by Lehigh County, and its seven board members are appointed by the County Executive for staggered five-year terms. LCA does not have any taxing authority, but it does have the ability to issue its own debt. All of its revenues are received for the provision of utility services. LCA's 2006 annual operating budget is approximately $10.3 million, which is almost balanced between wastewater and, water service.

LCA operates under the supervision of a General Manager. The General Manager oversees an Assistant General Manager, administrative support personnel, and three major functional groups: Operations & Maintenance, Administrative Services, and Capital Works.

A recent feasibility study for LCA found LCA to be "very strong financially" with "substantial reserves for both operational and future capital needs." LCA has managed to do this while having rates for service that are "among the lowest in the Lehigh Valley region" while still generating "substantial operating and debt coverage annually" (EMA, Inc. 2004).

Factors Contributing to Success

LCA has been part of the community for nearly 40 years and it has a reputation for providing high-quality water and wastewater services and for having a very professional, highly skilled staff. It remains highly visible in the greater Allentown area and works with other regional organizations, such as the Chamber of Commerce, the Lehigh Valley Partnership (an economic development partnership), and others.

LCA has developed an understanding of the neighboring utilities and has approached many of them individually to explore the possibility of joint efforts, ranging from cooperative activities (e.g., joint purchasing) and including but not limited to possible acquisition. While not all of these advances have been successful, these contacts strengthen LCA's position as a regional resource for smaller utility systems.

It is also noteworthy that even though LCA is not a large organization (approximately 35 employees), it has developed the in-house expertise to evaluate and ultimately consummate transactions with other utilities. This includes the engineering, legal, and financial expertise that is necessary to properly evaluate growth opportunities. Having this expertise in-house greatly reduces LCA's costs and improves its ability to move quickly, while assuring senior management and the Board of Directors that a potential growth opportunity will benefit existing customers and the organization as a whole.

Obstacles

All of LCA's successful acquisitions since 2000 have been of privately owned water or wastewater utilities. While LCA has approached publicly owned systems, acquiring a publicly owned system has proven to be more difficult because of municipal officials' concerns with losing control over their infrastructure. In some areas, the water and wastewater systems are a de facto method to control zoning and land use in the community. Local officials are reluctant to cede control of this infrastructure to an outside entity, even one like LCA that is a governmental entity.

In addition, while any community would welcome a one-time cash payment for its water system, local officials also must balance that against having to pay for water and wastewater service. For example, in some communities, a significant obstacle would be the cost of having to pay LCA's charges for public fire hydrants.

Some communities also view a loss of their utility systems as having an effect on other municipal services. It is not uncommon for the utilities to share employees and equipment (such as a backhoe or truck) with other municipal departments. Because of this sharing, the sale of the municipal utility will eliminate the revenue from the utility, but may not eliminate all of the expenses that are currently included in the utility's budget.

Similarly, some communities use their utility rates to subsidize other municipal services, keeping property taxes lower than they would be otherwise. As a result, the sale of the utility might lead to a need to increase property taxes. This is not only politically unpopular, but it also negates some of the benefit that the municipality otherwise would receive from the proceeds of the sale.

LCA also has noticed that some water and wastewater systems are not cognizant of, or do not fully appreciate, the risks of continuing ownership of their systems. For example, a municipality may view the water utility as a source of positive cash flow. Yet it may fail to appreciate that there is a significant risk of that cash flow turning negative if there is a substantial main break, a significant new treatment requirement, or the loss of a major customer.

Finally, the Pennsylvania governing system, in which considerable autonomy is granted to the smallest local units of government (e.g., townships), while little authority is ceded to county-level entities, creates another obstacle to regional approaches. The decentralized nature of local government authorities in the state creates barriers to county-level planning and leverage that might otherwise support more regional approaches to water supply provision.

Conclusions

LCA is building a successful, regional water and wastewater utility through a combination of internal growth, acquisitions of neighboring utilities, and regional partnerships. LCA recognizes that continued, cost-effective growth can help to keep its customers' rates stable, while maximizing the utilization of its existing personnel and other resources.

LCA has established itself as a regional resource for other water and wastewater utilities. In addition, it works closely with other regional organizations, such as planning and economic development agencies, further solidifying its role as a source of water and wastewater expertise within the region.

LCA's continued growth and success in improving the provision of service throughout its region can be attributed to several factors, including:

- Professional and highly skilled personnel
- Sound financial management
- Development and implementation of a strategic plan
- Longevity in the community, including a history of providing high-quality service at reasonable rates
- Working to build partnerships throughout the region, not just for utility service, but also for planning, economic development, and community improvement

MASSACHUSETTS WATER RESOURCES AUTHORITY (MWRA)

Background

The MWRA was established in 1984 and supplies water and wastewater services on a wholesale basis to 61 communities serving 2.5 million people and over 5,500 major industrial customers in eastern Massachusetts, including the City of Boston (MWRA 2005). The system's main sources of supply are the very large Quabbin Reservoir (412 billion gallon capacity) and the smaller Wachusett Reservoir (65 billion gallon capacity) in the central part of the state. The MWRA also manages four large covered distribution storage facilities with a total storage capacity of 180 million gallons (MWRA 2005). The safe yield of the system is approximately 300 mgd. Average water demand is about 230 mgd. Wastewater from member communities is collected by local sewer systems and then flows, mostly by gravity, to the recently reconstructed Deer Island Treatment Plant. This facility provides primary and secondary treatment and discharges the effluent into the Massachusetts Bay. On average, 350 million gallons of sewage are treated per day with a maximum flow of 1.2 billion gallons per day during extreme wet weather (MWRA 2005).*

Preceding the formation of the MWRA, Boston endured a long history of water governance marked by transitions between various authorities. Boston's regional water system dates back to the 1640s when a single cistern served 208 households (Wallace, Floyd Associates 1984). By 1795, the need for more water led to the creation of a log pipe system from Jamaica Pond; one of the first examples of taking water from an outside community to serve the Boston area (Wallace, Floyd Associates 1984). 1848 marked the development of Lake Cochituate as a water supply, as well as the first public undertaking of water management by the Cochituate Water Board. In 1895, the State Board of Public Health recommended the establishment of the Metropolitan Water District, the first metropolitan water supply organization in the Boston area.

At the same time, the Board of Public Health launched a study to assess water use by population sector, determine the amount of water leakage occurring, and identify alternative water supply sources to meet the rapidly growing demand for water. Of paramount concern to the Board were issues of water quality, specifically the contamination of water sources by sewage, and water quantity, with regard to population projections and increasing water use per capita (Wallace, Floyd Associates 1984). By 1907, a massive water project had already been completed to meet growing demand: the creation of the Wachusett Reservoir and Dam (Bowers and Carolan 1985). In the following 10 years, the need to construct new facilities declined and the focus shifted to the consolidation of the administrative aspects of the regional water system (Bowers and Carolan 1985).

Motivation for Considering Regional Solutions

Regionalism was enticing as a water and sewage system in Boston for a number of reasons. First, consolidation of water and sewage resources would reduce competition among municipalities for natural resources as well as combine financial resources for costly infrastructure construction. Before a regional system was adopted, municipal efforts to create individual water systems were stymied by competition with one another (Elkind 1998, p. 94). Second, the creation of a regional water system was attractive to many different groups in the political realm. The planning of large scale projects involved technical expertise and scientific innovation that was appealing to many. In addition, the creation of jobs from construction projects appealed to constituents. Those wary of government authority were likewise tempted by a regional system that would limit the authority of municipal governments. At the same time, participants in municipal governments were in favor of a regional water supply because it did not require a change in the existing local government structures. "Institutional innovations appealed to political reformers who wanted to limit municipal power. Meanwhile, the fact that special districts neither had direct oversight over existing city institutions nor required cities to abandon or transform their traditional governmental structures met with approval from municipal officials" (Elkind 1998, p. 80).

While many were in favor of a regional system, an element of skepticism lingered. In particular, residents were suspicious of new tax schemes and the abandonment of revenue-generating local water systems. Rural inhabitants were particularly wary of intrusion as a broader system

---

* There are three major reasons why, on average, MWRA treats more wastewater per day than it provides drinking water: (1) the water and wastewater service areas are not fully overlapping; (2) there is considerable inflow and infiltration into the system; and (3) several communities supplement MWRA water with local supplies, but discharge all of their wastewater to the MWRA system.

might reach farther west for new water supplies. Nevertheless, the general approval of a regional system by many different groups as well as the increasing demand for water account for the persistence with which Boston and surrounding communities moved toward a regional water system.

Two studies were instrumental in fueling the motivation towards regional solutions to water management issues: (1) the aforementioned study issued by the State Board of Public Health in 1895; and (2) a 1922 report, requested by the Massachusetts Legislature, from a Joint Board comprised of affiliates of the recently formed Metropolitan District Commission (MDC) and the Board of Health.

The Joint Board report concluded that the existing water system was insufficient due to expanding population in the area as well as increasing water use per capita due to improved standards of living and industrial growth. The Board estimated that within a decade the existing system would be unable to meet the region's water demands. To address the projected shortfall, the Board recommended diverting high flows (flood skimming) of the Ware River to the Wachusett Reservoir as well as the creation of a very large new reservoir on the Swift River. The new reservoir was proposed despite the fact that it would require the inundation of four towns and the relocation of all residents and businesses from these communities.

While these recommendations met with opposition and a new committee was created to develop alternatives, the potential inadequacy of existing water supplies remained at the forefront of managerial concerns. Though both population and water consumption projections issued by the two studies were higher than actual growth, water quantity remained the primary motivating force for cultivating and refining regional water systems. Furthermore, although the problem of increasing demand was the more evident and pressing problem, issues of water contamination complicated quantity problems because as demand increased, communities were reluctant to abandon contaminated water supplies. For example, Boston continued to tap Mystic Lake as a water supply even after the Boston Water Board and State Board of Health condemned the source (Elkind 1998, p. 95).

In response, construction of the Quabbin Reservoir began in 1936 and the reservoir was filled with water from the Swift and Ware rivers beginning in 1939. The Quabbin was filled to capacity in 1946 and solved many water quantity problems due to its tremendous storage capacity. Demands on the regional water system continued to grow for several decades, and in the early 1980s the MDC began to look for additional sources of supply, including the Connecticut River. With the establishment of the MWRA in the mid-1980s by an act of the Legislature (see below), there was a fundamental shift in perspective from one focused almost exclusively on finding new supplies to meet what were perceived to be ever growing demands,* to a more balanced integrated resources planning approach that emphasized demand management and improved system management. The MWRA quickly demonstrated that with significant investment in demand management programs its existing water supply in the Quabbin Reservoir was adequate to meet the region's long term demands, and Authority planners and others continue to believe it will remain a sufficient water supply source for the Boston region (MWRA 2005).

* The anticipated loss of local water supplies in communities partially supplied by MWRA was a major driver of the perceived need for additional MWRA supply.

What Happened: The Road to Success or Failure

Health concerns and a growing demand for water led to further regionalization of the water system. The creation of the Metropolitan Water District was the first step in the regionalization of water management and was followed at the turn of the century by the consolidation of the water and sewage administrations into the Metropolitan Water and Sewage Board (Bowers and Carolan 1985). In 1919, a legislative act created the MDC uniting authority over metropolitan water, sewage, and parks within one organization (Wallace, Floyd Associates 1984).

The creation of a regional water and sewage system in the Boston metropolitan region solved some pressing problems for Boston-area communities, most importantly meeting the water needs of residents and businesses. However, regionalizing the system also created some challenges along the way, which ultimately the regional system has had to address.

Early Failures

After some success in the decades following its establishment, expectations and circumstances changed after World War II and it became clear that the framework of the MDC was imperfect for structural, financial and managerial reasons. Much of the problem stemmed from the fact that the MDC was the creation of the state and had little independent authority outside of the state legislature. The rigid structure of the MDC could not be changed without legislative approval and therefore the MDC was unable to act outside of certain guidelines with respect to issues demanding regional cooperation (Wakstein 1972). In addition, throughout the MDC's existence, as the metropolitan region grew the service area for the regional water and sewer systems was expanded beyond the boundaries laid out in the initial legislative framework (Wakstein 1972).

In terms of financing, MDC funding was dependent on legislative appropriations and it did not oversee its own finances. Consequently, the MDC often lacked sufficient funds and was unable to provide adequate maintenance of its infrastructure. In particular, financial difficulties resulted in disregard for the wastewater treatment plants, which led to sewer overflows, contamination of the Boston Harbor, and a failure to comply with state and federal environmental regulations (AwwaRF 1997). By the mid 1980s, Boston Harbor had become known as one of the most polluted Harbors in the country. In response, in 1983 the Conservation Law Foundation filed a lawsuit in Federal Court against the MDC and the USEPA to force the necessary system repairs and upgrades, to restore Boston Harbor, and to bring the system into compliance with the Clean Water Act (CWA). The lawsuit was successful and in December 1984, the Massachusetts Legislature created a new water governing body, the MWRA (Conservation Law Foundation 2005). The MWRA acquired most of the water and wastewater assets and responsibilities of the MDC. In addition, the Legislature granted the MWRA the authority to raise money through revenue bonds and set rates for its member communities. No longer would the water supply authority lack sufficient funds to operate or to deal with contamination problems due to unstable and inadequate legislative appropriations (Conservation Law Foundation 2005). At the same time, the political problems of siting major new facilities and raising rates to pay for them shifted from the Legislature to the Authority Board.

From a managerial perspective, because the MDC was part of the state government, all appointments came from the state with little input from the communities served by the agency (Wakstein 1972). As a result, member communities and their residents over time viewed the MDC as an extension of State Government, a source of patronage hires, and unresponsive to member communities' needs. Also, the MDC's responsibility for parks and recreational facilities, which was so successful in the early years, became a distraction. These more visible parts of the agency took greater priority, and maintenance of water and wastewater infrastructure suffered.

Later Successes

While the MWRA is accountable to several governmental bodies, including the Federal Court, State Inspector General, and State Auditor, and must adhere to standards laid out by the CWA and the SDWA, the MWRA is an independent public authority. Furthermore, the MWRA is structured to avoid some of the structural, financial, and managerial problems that plagued the MDC.

In contrast to the MDC, the enabling legislation for the MWRA (Chapter 372, Acts of 1984) established an 11-member Board of Directors, eight of whom are directly or indirectly appointed by the 61 customer communities. Three members are appointed by the Governor, with two of them representing the river basins where the source reservoirs are located. The composition of the Board, along with the creation of several advisory committees, has resulted in an agency that is far more responsive to its member communities than its predecessor.

The Enabling Act also created the MWRA Advisory Board, made up of the chief elected official (or designee) from each of the 61 cities and towns served. The Advisory Board elects from among its members three of the 11 Board members, has statutory review power over MWRA's capital and operating budgets (but not approval), approves new additions to the water or sewer service area, and serves as a conduit for member community concerns. MWRA also voluntarily provides financial support to the independent Water Supply Citizens Advisory Committee (WSCAC) and the Wastewater Advisory Committee (WAC). WSCAC pays particular attention to protecting the interests of the source water communities in central Massachusetts. In carrying out its mission it performs technical reviews of projects and provides input into water supply, expansion, and other MWRA policy issues.

The success of these Advisory Committees derives from the MWRA's distinct managerial framework and a recognition that to run a successful regional system the needs and interests of multiple stakeholders need to be balanced.

In the financial sphere, though there is oversight from the Board and various advisory bodies, the MWRA manages its own finances and has the authority to create budgets, enter into contracts, incur debt, and issue bonds. This financial independence, along with the transparency created by the Advisory Board through review of the annual budgets, have been critical in the Authority's efforts over the past two decades to respond to the federal court's rulings and clean up the contamination in Boston Harbor. Moreover, it has provided MWRA with a reliable flow of resources and allowed the agency to do long-term capital planning and address long-neglected maintenance needs. Since its inception in 1984, the Authority has undertaken major improvements in both the water and wastewater system. For the water system this has included the construction of new water treatment plants, implementation of a watershed protection program, construction of the Metro West Supply Tunnel (a critical project for water transmission redundancy that was put on hold more than 50 years ago), and many others. Wastewater system improvements have included significant rehabilitation of much of the region's sewage infrastructure, completion of the $3.8 billion Deer Island sewage treatment plant, construction of a sludge-to-fertilizer plant, and the installation of a new treated wastewater outfall in Boston Harbor. Virtually all of these improvements were regional in nature and required the resources of all member communities of the regional system.

Supply and Demand

The MWRA differs from the MDC and in fact all other pre-existing Boston metropolitan water supply management bodies (and until recently, most water supply agencies across the nation) with respect to supply and demand management. The original studies published by the Board of Health and Joint Board in 1895 and 1922, respectively, alluded to demand management issues including leakage, metering, water rates, and consumer conservation. However, neither of these studies suggested demand management as an alternative to increasing available water supplies (Wallace, Floyd Associates 1984).

Likewise, the MDC focused heavily on supply management techniques and developing alternative supply sources when necessary. Conversely, the management scheme adopted by the MWRA focuses on demand management, primarily in decreasing losses to the water system, protecting local sources from contamination, and regarding the addition of supply sources as a secondary consideration. To supplement its demand-based management scheme, in the early 1990s MWRA adopted Trigger Planning as a method to determine if and when supply sources need to be supplemented.*

MWRA's focus on demand management and watershed protection has also addressed water quality concerns that have long-plagued the region. The Authority's efforts in this regard are guided by law. To ensure that the MWRA does not neglect or contribute to the abandonment of local sources, the Massachusetts Interbasin Transfer Act requires that for significant water transfers, the Massachusetts Water Resources Commission (MWRC, a state entity as opposed to the independent MWRA) must "determine that all practical water conservation measure have been taken and all viable local water sources have been developed in the receiving basin, and that reasonable instream flows will be maintained in the donor basin. Proponents should consider interbasin diversions to be a last resort after all other in-basin options have been exhausted" (Massachusetts Office of Coastal Zone Management 2005). Thus, the MWRA has adopted demand management as a way to address issues relating to both water quantity and quality.

In addition, the MWRA's effort to place demand management on equal footing with supply management balances the interests of source water communities in the central part of the state with those of urban populations in the Boston area of eastern Massachusetts (MWRA 2002, 2004). Many rural communities felt that their livelihoods were disrupted for the purpose of channeling water to a distant city. For example, after the Metropolitan Water Board had created the Wachusett Reservoir, a citizen of the town of Clinton declared, "No other public work so directly and inevitably involves injuring one portion of the State for the sole and exclusive benefit of another" (Elkind 1998, p. 113). Thus, while the creation of a regional water system diminished some of the conflict and competition between adjacent metropolitan communities, the system allowed a conflict to persist between rural towns to the west and urban communities in the east. Such conflicts have been reduced through MWRA's adoption of demand management schemes as well as prominent roles for representatives of source water communities on various advisory boards and the MWRA's Board of Directors.

---

* According to the Summary Report of MWRA's Demand Management Program for Fiscal Year 2003, under trigger planning, the first step is to identify "leading indicators" that can be monitored over time and act to "trigger" a response by the Authority. The second step is to analyze in advance to reduce the time for implementing projects. Other aspects include ongoing analysis of water use trends, as well as economic and demographic conditions and periodic demand projections.

Lessons Learned

Several lessons may be drawn from the experience of the MWRA and its predecessors in creating and operating a successful regional water (and wastewater) system. These lessons relate to institutional, financial, and political factors. Institutionally, the Authority's predecessor, the MDC, lacked the legitimacy that comes with a representative management structure. The composition of the MWRA Board of Directors and advisory committees reflects a conscious attempt to balance the interests of various stakeholders, including the rural source water communities in central Massachusetts and the more urban, consuming communities in the Boston area.

The enabling legislation that created the MWRA recognized the importance of financial autonomy and provided the agency with the requisite authority to raise sufficient resources, from both water and sewer revenue as well as from debt, to undertake major system improvements. At the same time, the enabling legislation and subsequent actions have led to a system of strong oversight and accountability to the member communities and ratepayers. While MWRA's water and sewer rates have increased dramatically, particularly during the Authority's first decade, accomplishments over the past 20 years are impressive. New water and wastewater treatment plants have been constructed and are now operating, as are many other major capital improvements. Though there has been tremendous economic growth in the region during this period and several new communities have become members of the system, through a variety of demand management programs and system infrastructure improvements, total average water demand on the system has decreased from over 330 mgd (which exceeded the system's safe yield) to about 230 mgd. And importantly, water quality and the overall ecological health of Boston Harbor have improved dramatically.

It is unclear whether the Legislature would have ceded such authority to an independent entity if there had not been a court order to comply with water quality regulations and clean up Boston Harbor. As noted above, by moving toward an independent agency the Legislature was able to avoid the need to site and fund controversial new system improvements, yet retained the option to criticize the MWRA's actions. Nonetheless, the experience of the MWRA as an independent body providing wholesale water and wastewater services implies that its structure may have qualities that could serve as a model for other regional systems. Characteristics such as broad stakeholder representation in decision-making, close coordination rather than competition with local municipal systems, financial autonomy, and integrated consideration of both demand and supply options are all elements that may be usefully transferable to systems in other regions.

METROPOLITAN WATER DISTRICT OF SOUTHERN CALIFORNIA

Background

The MWD was formed on December 6, 1928, under the Metropolitan Water District Act, for the purpose of transporting water from the Colorado River (Vaux 1988). It serves 26 cities and water districts in an area of 5,200 square miles in parts of Los Angeles, Orange, San Diego, Riverside, San Bernardino, and Ventura counties. An average of 1.7 billion gallons of water per day is distributed to 18 million people, making MWD the largest urban water supplier in the U.S. The two principal sources of water are the Colorado River (delivered via the Colorado River Aqueduct) and the Sacramento-San Joaquin River Delta (delivered via the Edmund G. Brown California Aqueduct). MWD's infrastructure includes 775 miles of distribution pipelines, 6 pumping plants, 5 water filtration plants, and 16 hydroelectric plants, allowing for nearly 1.1 million acre feet of storage capacity (MWD 2005e).

MWD is a water importer, serves its members through wholesale contracts, and supports and assists in the development of local water supply resources (Baumann, Boland, and Hanemann 1997). MWD provides its member communities with additional water rights and the capital with which to develop these additional supplies. The percentage of MWD water used by its member agencies ranges from 0% (San Fernando) to 100% (Las Virgenes Municipal Water District) (MWD 2005f).

MWD is incorporated as a local unit of government and requires electoral approval by all its member municipalities. MWD is governed by a board of directors, consisting of 37 members, representing all member agencies, and appointed by the chief executive officers of the various member agencies (Vaux 1988, MWD 2005e). There is one director vote for each $10 million of assessed valuation, and every member agency is entitled to at least one vote and none can have more than 50% of all votes (Vaux 1988). The total assessed valuation of MWD member agencies is $1.625 trillion, and the largest share, $308 billion, belongs to Los Angeles, giving them 19% of the total vote (Means and Malinowski 2005).

Motivation for Regional Solutions

As early as the 1880s, Southern California experienced explosive population growth. From 1880 to 1905, for example, the population of Los Angeles grew from 11,183 to 250,000 (Vaux 1988). As this population influx continued, pressure on the water supply became a growing concern. The mean annual precipitation of Southern California is 14 inches a year. However, in the early twentieth century, Southern California experienced an 11 year drought in which annual precipitation was as little as 40% of that (Vaux 1988). The drought coupled with the explosive population growth made it apparent that the predominant supply of domestic water, the Los Angeles River, was less than adequate.

There were two major reasons for the creation of the MWD: the need for an adequate supply of water, and political concern for local control. In 1913, the Los Angeles City Water Department constructed an aqueduct to transport water from the Owens Valley, but the demand for water grew faster than the Owens Aqueduct could sustain. Other municipalities in the region were also running low on water. For example, Anaheim's wells were being drilled to 600 feet in mid-1920s, compared to 15 feet in late 1800s (MWD 2005b). At the same time, Southern California was at the height of the political controversy, now dubbed the California Water Wars, in which Los Angeles was accused of "forcing" the annexation of smaller municipalities in the region so that they could obtain water from the Los Angeles River (MWD 2005b). As Los Angeles and the other districts of Southern California grew more and more desperate for an adequate water supply, advocacy grew to find a mechanism to allow use of water from the Colorado River, which would require large capital outlays to build the importing infrastructure.

The regional water needs of Southern California spanned numerous local political jurisdictions and required more financial (and political) capital than any one municipality could afford. Therefore, the state created an independent agency, MWD, to develop and distribute the water of the Colorado River while protecting the independence of smaller municipalities from the political powers of Los Angeles (MWD 2005b, Vaux 1988). MWD board member, Timothy Brick, describes the need for a metropolitan water district as follows: "Three problems faced the visionaries [of the Colorado River project]: the cities were not contiguous, they lay in four different counties, and the smaller cities needed safeguards to protect them from Los Angeles' enormous wealth and power. The solution they devised was the creation of a special district authorized by state legislation, a metropolitan water district" (MWD 2005b).

What Happened: The Road to Success or Failure

In 1931, MWD obtained a bond for $220 million for the construction of the Colorado River Aqueduct, and Colorado River water was being used by the MWD's service area by June, 1941 (Vaux 1988). MWD was given the fourth-priority right of California's share of Colorado River water, which allowed them 550,000 acre-feet of water per year. In addition, the City and County of San Diego gave the MWD their fifth-priority right of 662,000 acre-feet (MWD 2005d).

In 1988, MWD formed a partnership with the Imperial Irrigation District, IID. This partnership is known as the IID/MWD Water Conservation Program. MWD provided the funding for conservation improvements to the IID's distribution system. The project involved the construction of concrete lining earthen canals, local reservoirs, and spill-inceptor canals; installation of non-leak gates; automation of the distribution system; and altering delivery timetables. The conservation measures have resulted in 106,880 additional acre-feet of water per year and are expected to continue through 2033 (MWD 2005f). This is another example of a large-scale regional approach wherein different water agencies worked cooperatively to find a mutually beneficial manner to better manage scarce water resources in the region.

In 1993, MWD engaged in an Integrated Resource Planning (IRP) effort aimed at correcting two related, fundamental issues: lack of coordination between MWD and local agencies and amongst the local agencies, themselves; and the limited supply of water. The local agencies had developed independent strategies to cope with the region's limited supply of water. These strategies, however, were not always compatible with one another. Through a work group, focus groups, public forums, and regional assemblies, MWD was able to regionalize a coordinated resource strategy. After comparing the marginal benefits with the marginal costs of reliability it was decided that an optimal level of reliability was preferred to absolute reliability. MWD thus created a diversified water resource portfolio for managing the region-wide needs, consisting of 22% conservation, 22% State Water Project, 18% Colorado Aqueduct, 16% storage and transfers, 13% groundwater storage, and 10% recycling and reuse (Baumann, Boland, and Hanemann 1997).

The distribution of votes between the member agencies of MWD has raised some concerns. The distribution of votes is based on assessed valuation rather than water usage. This creates an imbalance between resource reliance and representation. For example, San Diego County relies on MWD for 95% of its water (via the San Diego County Water Authority). It accounts for one forth of all MWD sales, yet is allocated only 15.5% of MWD's board votes and has only 13% preferential right to capacity (Brackman and Erie 2001). This imbalance has motivated San Diego to desire and gain some independence from MWD (Brackman and Erie 2001). Similarly, the Las Virgenes Municipal Water District is 100% reliant on MWD for its water and receives only 0.94% of the votes; Los Angeles, on the other hand, is only 30% reliant on MWD for its water supply yet holds the largest allocation of votes at 19% (Means and Malinowski 2005, MWD 2005f).

Environment and Conservation

Through 1970, MWD acted with little regard to the environmental sustainability of their water supply. But as Martha Davis, former Executive Director of the Mono Lake Committee, points out in her speech to the UCLA Environment Symposium, "If the State's first fifty years of water development was about the construction of dams and aquifers to meet LA's and California's growth needs, the second fifty years has been about coping with the environmental problems created by those projects" (Davis 1998).

During the 1987–1990 drought, MWD implemented water conservation programs, improved water recycling capabilities, and improved groundwater management in order to cope with the decreased supply of water. Today the MWD's service area uses the same amount of water as it did 15 years ago, despite a 30% population increase (MWD 2005c). In 2003, recycled water production was estimated at 204,000 acre-feet; Southern California water users as a whole conserve 480,000 acre-feet of water per year (MWD 2005a). In 2003, MWD committed $20 million to water conservation programs (MWD 2005a).

Conclusions

The member agencies of the MWD opted for a regional approach to water management due to the need to develop (import) an adequate water supply to meet growing demands that outpaced locally available supplies. A key concern driving the formation of MWD as an independent entity stemmed from political concern about local autonomy and the fear of undue influence of the largest and most powerful cities (e.g., Los Angeles). The District was set up as an independent, local entity of government to protect the independence of smaller municipalities and to increase the capability of raising funds for the needed infrastructure. The regional approach of Southern California has enabled increased water supply, improved coordination among both member and outside water agencies, and has allowed for improved water conservation measures. While frictions do exist over the distribution of power among the member agencies, and it may be difficult to manage an organization that represents 37 different entities, on the whole MWD has been a successful example of a regional solution to water supply.

NORTH CAROLINA CENTRAL COASTAL PLAIN

Background

The Central Coastal Plain (CCP) of North Carolina is a prime example of demand outstripping supply and illustrates many of the challenges and opportunities that are typical for regional approaches in water scarce regions. This section provides an abbreviated description of this case study, and a more extensive version (including empirical information on estimating the quantified benefits of regional approaches) is provided in Appendix B.

North Carolina is rife with small water systems, as over 90% of community systems serving populations of fewer than 3,300 people. The CCP is a rural region in the eastern third of North Carolina that is dotted with small towns. A group of formations collectively known as the Cretaceous Aquifers supply the vast majority of water to communities in the CCP. The aquifers have a reasonable yield and produce a high quality water that generally requires little, if any, treatment, resulting in an inexpensive water supply.

As population and water demand have increased in the CCP, water levels within the aquifer have been declining. The state has classified drawdown of a foot or more per year to be "significant," a rate that has been exceeded in many regions within the CCP. Declining water levels have resulted in two other related problems. Aquifer dewatering, which occurs when the water level drops below the aquifer's confining layer, can lead to strata compacting and the permanent loss of storage space within the aquifer. Salt water intrusion is also an issue, as a wedge of salt water (water with elevated dissolved solids levels) is moving inward from the coast.

Through out the 1980s and 1990s, monitoring data suggested that these problems were growing in severity and that the future viability of the formation was at risk. In response, the state invoked the Water Use Act of 1967, which authorizes the state to declare the affected region a Capacity Use Area (CUA). Once a CUA has been established, the state can promulgate rules that regulate the threatened resource to ensure sustainable usage. The Central Coastal Plain Capacity Use Area (CCPCUA) that was ultimately defined covers an area including all or parts of 15 counties.

Motivation for Exploring Regional Approaches

Because of the apparent adverse impact of over-pumping of the regional groundwater supply, the final CCPCUA rules aim for groundwater withdrawals to be reduced significantly over a three-phase, sixteen-year period (NC DENR 2000). Every groundwater user pumping more that 100,000 gallons per day (gpd) will be required to reduce groundwater consumption by anywhere from 10% to 25% (relative to current use) at the end of each of the three phases (6 years, 5 years, 5 years), resulting in total reductions between 30% and 75%. There is no accommodation provided for water systems whose demand rises over the 16-year period, so at the end of this time many users will be entitled to withdraw even less than 25% to 70% of their daily demands.

The reasons that make groundwater so attractive within the CCPCUA (low treatment costs, widespread availability) make surface water equally unattractive. There are several rivers within the region that could serve as water sources, but in practical terms, only two, the Neuse and Tar rivers, are thought of as viable water supplies. In addition, very little surface water treatment capacity in the region, with only one surface water treatment plant on each river (Goldsboro on the Neuse; Greenville on the Tar). Tapping these surface sources will require substantial investments, particularly given that (1) extensive surface water treatment would be needed, and (2) many communities do not lay along the rivers, so conveyance infrastructure will also be required to bring surface water to these groups.

In considering alternative strategies for meeting the water demand of these small communities, regionalized facilities began to appear attractive. Surface water treatment exhibits significant economies of scale, with the average cost of produced water decreasing rapidly as a plant's capacity increases from that of very small plants (< 1 mgd) to larger ones (around 10 mgd). The vast majority of affected communities in the CCP have demand for less than 2 mgd, as such both the North Carolina Department of Environment and Natural Resources (NC DENR) and local planning assistance agencies (e.g., North Carolina Rural Center), encouraged the affected communities to begin considering cooperative approaches to meeting regional potable water demand. Still, a number of challenges remained.

Challenges to and Options Considered for Regional Solutions

First among the challenges was simply convincing people there is a problem, because trying to convince local politicians that they need to spend considerable amounts of money on a problem that they cannot see with their own eyes is a difficult task. There were, however, areas in which problems had reached a point where they had become self-evident. Kinston is one of the larger cities within the region and the area in and around it has seen some of the largest declines in aquifer levels.

The challenge of making the pumping restrictions equitable fell to officials at NC DENR, Division of Water Resources, who developed an interesting structure for the cutbacks. One early lesson in developing these plans, however, was the need for transparency. Once the communities had a firm understanding of their status, they were able to begin an organized planning process to meet the impending pumping reductions.

In the case of a number of communities, these plans have included the formation of regional surface water systems, which include both regional treatment facilities and conveyance networks, with the increased scale of the facilities allowing for some reductions in per unit cost. Presently there are two different groups in the CCPCUA taking two separate paths to regionalization. The first group is the Neuse Regional Water and Sewer Authority (NRWASA) which will tap the Neuse River and is centered in Lenoir County. The second group is forming around the city of Greenville which draws water from the Tar River.

With the creation of the CCPCUA, the county commissioners of Lenoir County urged the five water systems in the county to form a single regional system. As of July 2005, plans for the treatment plant, to be located a few miles upstream on Kinston on the Neuse, are being prepared to go out to bid. Current expectations are that construction of these facilities will be completed in 2008, the same year in which the first phase of cutbacks are to be enacted. In this case, it made sense to build the system large enough to begin to move these communities entirely off of groundwater, giving each a significant volume of excess permitted pumping capacity. The size of the regional system has grown since its initial inception, as the original five Lenoir County members have more recently been joined by four additional water systems from outside the county. The state legislature has also sought to make more alternatives available through legislation allowing private, non-profit water corporations to partner with public water systems in efforts to form regional systems.

One of the biggest proponents of NRWASA has been Ralph Clark, Kinston's City Manager. Kinston is easily the largest water system in the NRWASA system and lies directly on the Neuse River. Consequently, Kinston would logically be in the best position to build its own, separate, surface water system, yet Clark sees NRWASA both as a necessity and a win-win situation for all involved. Clark claims that Kinston simply could not have afforded a surface water treatment plant outside of a regional scheme. "How much burden can you put upon 11,000 people?" he asks. As the largest city in NRWASA, Clark feels that Kinston's role is not to dictate terms to the other members, but to help provide the leadership to pull everything together in the end. This is a viewpoint that is easy to understand once it is recognized that every member contributes something. In the case of NRWASA, Kinston owns the permits to the Neuse River water, while several of the smaller systems have access to USDA assistance, programs not available to Kinston.

When asked about the challenges that have arisen as the system has developed, Clark says they are no different than any other situation in which one has to get four or five different groups to agree on something. Addressing this challenge forms the basis of the take home message that he often shares with others. "You have to tell everyone, 'It's a lot of giving and taking.' Every member is equally important." It is a realization that has served NRWASA well, as they have been asked to speak to other communities around the state facing similar groundwater issues.

Greenville has taken a different approach to creating a regional system. Greenville currently treats surface water from the Tar River to meet its demand, and as such, is not directly impacted by the CCPCUA regulations. It is, however, located in the heart of the region, and its 22.5 mgd surface water treatment plant has considerable excess capacity. After the CCPCUA regulations were passed, Greenville invited a number of local utilities to a meeting. The ensuing discussion revolved around Greenville's excess capacity and its willingness to sell wholesale water to those systems that were interested. Since then, two communities, Farmville and Greene County, have signed 40-year contracts in which Greenville agrees to provide them water on an interruptible basis. The agreement has Greenville supplying substantial quantities of water to these communities during periods when it has excess capacity, but it reserves the right to withhold that water up to 36 days per year when it faces peak demands in its own service area. As the groundwater cutbacks are based on a community's average annual consumption, Farmville and Greene County's permitted groundwater capacity following the cutbacks will be adequate to meet their demand when Greenville interrupts the service.

At present, all parties seem satisfied with this agreement. Greenville is able to use its excess capacity, while Farmville and Greene County are able to meet their future water demand in spite of the impending withdrawal reductions. Randy Emory, head of Greenville's water utility, cites very few problems in the process. There are some differences among the water systems, such as Greenville's use of chloramines and the groundwater systems' reliance on chlorine as a disinfectant, but plans for pipelines to connect the water systems are progressing. This arrangement has been so successful that other local water systems are now in discussion with Greenville about entering into similar contracts.

Factors Contributing to Anticipated Success

Throughout this process, water systems throughout the region have had several resources available to them. In addition to the NC DENR, the North Carolina Rural Center has worked to provide technical expertise and financial advice to communities facing the prospect of building new water-related infrastructure in the CCP. The Rural Center was established to address the needs of small communities, and has played an active role in assisting the affected communities. Jean Crews-Klein, Vice President of business and natural resource development, sees the role of the Rural Center as two-fold: facilitating communication among the parties and providing assistance in the funding process. Crews-Klein notes the challenges facing regional systems by saying, "The most pressing problem is where to get money for a regional system. Getting them started is a huge effort that requires huge capital investment. There is way more need than capital to address [the problem], and it places funders in a difficult position: Do you fund a system that may be having health problems, or do you fund a regional system because it's the right thing to do?" Nonetheless, she expresses pleasure at the amount of progress that has been made in the CCP, progress largely facilitated by thoughtful dialogue among regional cities since the pumping reduction rules were passed.

Another factor contributing to success is that the CCPCUA has spawned research that has explicitly examined the potential benefits of combining regional surface water systems with tradable groundwater permits as a means of cost-effectively coping with a reduction in groundwater pumping capacity (Kirsch and Characklis n.d.). Regional surface water systems would allow small systems to take advantage of economies of scale, while the sale of their groundwater permits would both offset the costs of joining a regional surface water treatment system and provide a means of augmenting supply for cities located further from surface water sources. In the case of the latter, joining the regional surface water system would be very costly, and it is often much less expensive for these communities to remain entirely on groundwater.

In the research effort, a model was created that evaluates both regional surface water systems and tradable groundwater permits as a means of meeting regional demand in the face of groundwater pumping reductions. The model's premise is that each community is an individual entity that acts in its own best interest and is responsible only for its own costs. The choice of water supply is based entirely upon cost, and the criterion that each city uses to compare water supply alternatives is the unit price increase at the tap ($ per thousand gallons). What necessitates a break from previous linear programming research is the fact that a city's cost for joining a regional surface water system is interdependent on the cost of groundwater permits. This results in a circular system in which increased groundwater permit prices lead to more cities joining regional surface water systems, but as more cities join regional surface water systems, demand for groundwater permits drops, driving their cost lower. In order to resolve these circularities, an equilibrium approach was developed to determine each city's least expensive water supply choice given the choices made by every other city in the region.

The results of the modeling research reveal the magnitude of cost savings that may be realized by regional approaches. Absent regional solutions, the total present value costs for a 30-year time period across the region would be over $252 million. Enabling regional surface water approaches such as described above, but restricting the groundwater permit trading, the present value cost decreases by nearly 25%, to $191 million. By concurrently enabling tradable groundwater pumping permits in concert with regionalized surface water treatment, an additional cost savings of $27 million can be realized (the present value cost would be $164 million), resulting in a combined 35% cost savings relative to the no regional solutions approach. Thus, the modeling research reveals that cost savings on the order of 25% to 35% may be realized, if regional approaches are adopted in this 15-county area. (Additional detail on the conceptual approach used, and the empirical results, are provided in the more extensive version of this case study, found in Appendix B).

As the empirical research model results suggest, regional surface water systems can provide communities with attractive alternatives that can significantly lower the costs of surface water treatment. The economies of scale that regional systems gain are a strong incentive for their creation. In addition, the development of a market for trading groundwater pumping capacity can provide communities with greater flexibility, and even lower costs, as they seek to meet their water supply needs.

Lessons Learned

The process of establishing the CCPCUA and working to meet its requirements has had its difficulties, but the mood of most everyone involved is now one of optimism. Numerous lessons have arisen out of the experiences of the CCPCUA, but as progress continues the overarching lesson from both NRWASA and the Greenville system is that regional approaches can work:

- Regional solutions allow communities to each contribute something and to play to their strengths. Larger cities may have better bond ratings than smaller communities, but the smaller communities may have access to funding sources (e.g., USDA) that are not available to larger systems.
- Taking advantage of economies of scale will be critical in the development of regional systems. While the planning and negotiation associated with developing regional systems will always take time, this approach can result in substantial reductions in unit costs, particularly for small communities.
- The use of markets, such as with the tradable groundwater permits, may provide an efficient means of distributing water rights throughout a region and benefit both buyers and sellers.
- Communication is key at every step of the planning process. Communities must know exactly where they stand with respect to regulators as well as potential partners in any regional scheme.
- There need not be a one-size-fits-all regional solution (e.g., Greenville vs. NRWASA). Ultimately, the creativity of the involved communities will provide them with the best solution to meet their needs.

PORTLAND, OREGON

Background

Water utilities in the Portland, Ore. metropolitan area have worked together over the last decade to advance a regional approach to water purveyance and management, as population growth and ecologic issues have placed greater stresses on available water resources. In this three county area, thirty-eight agencies provide water services to about 1.3 million people. The City of Portland's Water Bureau operates the water supply system that serves the majority (786,700) of these people, of which 482,500 are Portland municipal customers. Portland supplies water to the remaining 300,000 through wholesale contracts to 21 water providers.

The city's primary source of supply is the Bull Run Watershed, an area comprising 102 square miles of federally owned and protected land. Infrastructure in the watershed consists of two reservoirs (with 10 billion gallons of usable storage), a chlorine injection system, and the headworks to the conduit system that transmits the water by gravity to town.

The Bull Run watershed typically experiences enough rainfall to keep the reservoirs full for 8 to 10 months of the year. It is only during summer that inflows are reduced and the capacity of the reservoirs is relied on to meet in-town demands. This also corresponds with the period in which demands are at their peak due to warm weather and landscape irrigation needs. For these reasons, most planning activities, with respect to infrastructure sizing and conservation programs, are focused on this "drawdown" period during which the reservoirs are not full.

Type of Regional Approach Explored

As noted above, the regional approach historically in effect in the Portland area is one in which the city's Water Bureau is a wholesale provider to numerous neighboring communities. This case study examines a recent exploration by the region's various water agencies to explore a more integrated approach of shared ownership and responsibilities.

Since 1996, the Portland Water Bureau and a number of other local water providers have explored cooperation through the "Regional Water Providers Consortium," an entity formed by intergovernmental agreement. The Consortium, comprising 23 water providers in the Portland metro-area, serves as a forum for cooperation on water supply planning, resource management, and conservation issues.

In 2001, this dialogue evolved into a more concerted effort to create a formal regional water agency as many of the existing long-term contracts with the wholesale providers were approaching renewal. At this time, Eric Sten, the Portland City Commissioner with oversight responsibility for the Water Bureau, saw the opportunity to replace existing wholesale contractual relationships with a cooperative structure based on mutual ownership and stewardship of the area's water resources, including the Bull Run Watershed. The possible structure and composition of this regional water agency was debated and explored in two phases over an eighteen month period in 2001–2002, as the Water Bureau and 13 other local water providers came together to document their needs and wishes for such an entity.

The six-month long first phase of this exploratory effort began with several public meetings in which the regionalization effort was outlined and local water utilities' interest in creating a regional agency was gauged. This was followed by a series of weekly meetings in which senior managers from each of the participating utilities outlined possible governance models for the agency and synthesized common interests into a set of criteria that would serve as guidelines for formation of the regional agency. Phase I culminated in the identification of thirteen partner utilities, including the Portland Water Bureau, willing to commit to further pursuing the possible creation of this regional water agency.

During the second phase, which occurred during the first nine months of 2002, the thirteen partner utilities formed small "work groups" to analyze in greater detail the engineering, financial, and legal/governance issues that needed to be addressed and to amend the original set of organization criteria to reflect the input and concerns of local governments and private citizenry. From these analyses and input, a Principles of Agreement was drafted among the partnering utilities; this agreement was to form the basis of the new agency.

Ultimately, this 2-phased process of discovery did not culminate in the establishment of the envisioned regional agency. Within the span of several months, the initiative died due to a lack of resolution on several key issues identified during the two exploratory phases. These issues could not be addressed to the satisfaction of all the partnering utilities. In its place, talks began to renegotiate long-term contracts between the CoP and its wholesale customers. The discussion below provides an abbreviated assessment of the change in regional approach that was considered in the Portland area, and greater detail is provided in an expanded quantitative case study, provided in Appendix A.

Key Motivations and Factors Contributing to the Potential Success

Representatives from utilities that participated in the regionalization effort were asked to develop a list of the factors that most motivated them to engage in this initiative. One of the most cited motivations for participating in the regionalization effort was the opportunity for shared ownership in a supply source. Many of the partnering utilities that were existing wholesale customers placed great value on gaining an ownership role and the accompanying ability to participate in decision-making on system improvements and priorities. From the Portland Water

Bureau's perspective, the shared funding burden of capital improvements and expansions was also an important component to this ownership role.

After ownership, the most cited drivers for participation were the opportunity to conduct more efficient supply planning and the potential for cost savings through economies of scale (Portland noted that economy of scale savings may occur mostly through the water distribution operations). These savings would be manifest in lower long term rates for the metro-area customers. Portland also acknowledged that establishing a known customer base and supply through the regional agency could substantially enhance the efficiency of the planning process, as well as demonstrate the continued importance of the Bull Run system to the region's water needs, further ensuring its unique protected status into the future.

Other motivating factors relate specifically to water supply or governance issues. These drivers include the opportunity to diversify water sources (in reference to reducing risks associated with shortages or interruptions that may strike one source or type of source), to provide supplemental sources, and to diminish competition among the utilities for water. Other representatives cited the regionalization effort as an opportunity to create a good governance structure and bolster public opinion in government. A related response mentioned the fostering of a participatory structure as a driving force.

Thus, the goal of developing a more shared ownership and management of the region's water supply natural and capital resources (i.e., water and infrastructure) was envisioned as a way to promote more efficient planning and coordination, and thus realize a concomitant cost savings. However, the cost savings potential was never explored in any systematic, quantitative manner. Thus, the potential fiscal benefits were never made evident. Appendix A provides a more extensive discussion of supply reliability and investment timing issues that broader regional integration would influence positively.

In looking ahead to the future of water resource management in the Portland metropolitan area, there were a number of positive outcomes that stem from this aborted regionalization effort. Many of the partner utilities we talked with emphasized these outcomes as proof that, although the broad regional agency did not materialize, there will be lasting positive impacts for water management in the Portland metropolitan area that stem from this regionalization initiative.

Key Obstacles That Created Barriers to the Proposed Change in Regional Approach

Several interlinked factors appear to have ultimately prevented the implementation of the Principles of Agreement developed by the partner utilities. As discussed below, these factors relate largely to the evolution of the proposed agency scope; this scope eventually included only the Bull Run and Columbia South Shore well field supply system assets owned by the Portland Water Bureau. The initiative ultimately died trying to address the question of how, or if, the CoP could be adequately compensated for the transfer of these assets and the loss of the revenue potential of the Water Bureau, which would be transferred with these assets to the new agency.

From the potential partnering water agencies, the buy-in cost was seen as too high. In other words, the compensation to be paid to the CoP for the value of the Bull Run, Columbia well field, and transmission assets to be assigned to the regional agency was higher than they believed it should be. Many of the partnering utilities suggested they had already substantially compensated Portland for these assets through depreciation payments to the City made under their wholesale contracts over the prior three decades.

In discussions with partnering agencies, a belief was expressed that the process would have benefited from the use of a decision support system that would have allowed them to quantify and communicate particular supply and cost benefits that regionalization may have brought to the partnering utilities and Portland metro-area water customers. In our case study, we outline the value of such a decision support tool in this regionalization process, using WEAP (Water Evaluation and Planning system) to provide example demonstrations formulated from specific data acquired from the Portland metro-area water utilities.

The reluctance of the CoP to relinquish ownership of the Water Bureau and the Bull Run and well field assets because of their long-term revenue generating potential was also cited as a critical factor in ending the effort. Many believed the city had not fully comprehended this revenue factor when they formally supported the regionalization effort at the outset of Phase I, and that the economic downturn of the last two years had underscored the city's reliance on Water Bureau revenue that they believed could not be adequately compensated for by a regionalization "buy-in" amount that would be acceptable to all parties.

Also, a strategic error appears to have been not including key CoP stakeholders, particularly the Office of Financial Management, in the Phase I and II discussions, where the valuation question could have been handled earlier and with broader input from the CoP administration. This strategy had been put forward by the early champions of the effort, to minimize risks of the initiative getting bogged down by details and losing momentum. However, as the proposal moved forward into Phase II, the city's financial administration and the public began to view the initiative with concern. A lack of confidence in the Water Bureau that had developed during an ongoing billing system crisis grew to include the perception that the Bureau could not negotiate the regionalization process with the best interests of the city in mind. This lack of confidence in the Water Bureau was highlighted in late 2001 when the Mayor reassigned responsibility for the Water Bureau to Commissioner Dan Saltzman. This reassignment removed the original champion of the regionalization effort, Eric Sten, from a critical leadership role in the regionalization efforts at a critical point in the process.

The issues surrounding the valuation of the Bull Run assets were exacerbated by the city government's perception of the negotiated agency scope, which had evolved from that of a broad entity incorporating all pertinent water sources and distribution systems belonging to the partnering utilities to one comprising only CoP assets. Although the initial proposal suggested formation of an agency with broad control over regional supplies, transmission and distribution, partnering utilities indicated their desire early to retain control of their distribution and billing structures. From their perspective, the scope of the agency should center on regional water *supply and transmission* only.

Thus, the inclusion of individual utilities' distribution systems within the control of the agency was left as an *option* for the future. The distribution system owned by the Water Bureau continued to be considered for probable transfer, along with the Bureau, to the new agency. The city's government and Portland citizenry viewed this as an unbalanced agreement, with the smaller partnering utilities appearing to contribute little to the new entity while the CoP was giving up control of a substantial revenue-generating utility.

Lessons Learned

Partnering utilities were asked to reflect on the regionalization initiative with respect to what important lessons they felt were learned by conducting the process. Many also took this opportunity to contribute suggestions for different approaches that may have led to a successful outcome.

A number of utility representatives believe that if given another regionalization opportunity, they would engage critical stakeholders much earlier, particularly the staff of the Portland Office of Financial Management. Many representatives further stressed the lesson that strong driving forces and motivations have to be clearly communicated to the public and other stakeholders early on to build political and public momentum behind the initiative.

Representatives also indicated that having some sort of strong external driver for the regionalization, such as a regional water shortage, would have provided a driving force to push such an initiative to implementation. Simply put, the Portland metro-area public didn't perceive any compelling need to pursue regionalization.

The representatives also now realize that such an effort takes much longer to implement than they initially expected. They now see great future value in the engineering and financial data they acquired for and from the process. They admit, though, that they substantially underestimated what would be needed to plan such an undertaking. The partners believe that if they had known the length of time that would be required to pull together an agreement, they would have more effectively planned for navigating the process through the changing political and economic conditions that occurred and which had a substantial impact on their bargaining position with the CoP.

Finally, the partnering utilities stressed the importance of the revenue factor in the eventual decision to not continue efforts that may have led to formation of a Portland-area regional water agency. The CoP reacted very negatively to the potential loss of future revenues from transfer of the Water Bureau to the proposed regional agency. Because many city-owned water utilities provide substantial general revenue to the municipalities they serve, this factor could play an important role in other regionalization efforts across the country. From a policy and political perspective, foreseeing and addressing this key financial consideration could aid in ensuring a regionalization effort's success.

CENTRAL PUGET SOUND, WASHINGTON

Background

The Central Puget Sound Water Suppliers' Forum was formed in 1998, by 14 members. The founding members include water utilities for three cities (Tacoma, Seattle, and Everett), three counties (Pierce, King, and Snohomish counties), and 8 associations representing other water utilities in the region (Cascade Water Alliance, Pierce County Regional Water Association, North Snohomish County Water Utility Coordinating Committee, Everett Water Utility Committee, East King County Regional Water Association, South King County Regional Water Association, Snohomish River Regional Water Authority, and Water Suppliers Association).

The primary source of the region's water supply is the many rivers and creeks that drain the Cascade Mountain Range and flow through King, Pierce, and Snohomish counties. These surface waters carry runoff from precipitation, including delayed runoff captured in the annual snow pack. The other source of water supply in the region is groundwater. Groundwater supplies are primarily stored in unconsolidated sand and gravel aquifers derived from glacial deposits. In

Table 7.2
Public water supply in the Central Puget Sound area by source of supply

County	1995 total population served	% groundwater production	% surface water production
King	1,506,760	29	71
Pierce	580,550	66	34
Snohomish	429,650	16	84
Totals	2,516,960	37	63

Source: Based on USGS 1995

effect, these aquifers serve as an enormous "reservoir" for the region, by storing water that can be tapped for year-round use. Water users that rely on groundwater rather than surface water include many self-supplied commercial and industrial facilities, private households served by wells, and most of the small- to medium-sized water utilities that have independent supplies.

Approximately 63% of the three-county region's population rely on surface water as their primary drinking water source, and 37% utilized groundwater (USGS 1995). However, the use of surface water versus groundwater varies significantly by county. Table 7.2 indicates the population served and the percent of groundwater and surface water used for municipal water supply in King, Pierce, and Snohomish counties.

Approximately 63% of the three-county region's population rely on surface water as their primary drinking water source, and 37% utilized groundwater (USGS 1995). However, the use of surface water versus groundwater varies significantly by county. Table 13 indicates the population served and the percent of groundwater and surface water used for municipal water supply in King, Pierce, and Snohomish counties.

Motivation for Considering Regional Solutions

Though seemingly water rich, portions of the Central Puget Sound area face challenges in meeting existing and projected water needs for its people and fish habitat. The area has experienced significant growth over the past decade, with projected future increases. At the same time, the region is working to recover threatened species of fish by increasing in-stream flows and improving fish habitat. In this context, the Central Puget Sound Water Suppliers' Forum (the Forum) was formed.

The Forum is a cooperative regional effort to address current and future water supply issues, including water supply planning, environmental stewardship, and other issues currently facing the region. The Forum's vision is to develop sustainable water resource management that is responsive to (1) the Federal ESA, which could require certain levels of water quality or quantity in some waterbodies to support aquatic habitat, and (2) the Washington State Growth Management Act, which requires securing water rights for future land use planning efforts, while (3) recognizing tribal and stakeholder interests.

Recognizing Shared Challenges Helped Pave the Road to Success

The identified need for enhanced collaboration and communication between utilities arose as a result of discussions among utility leaders about the emerging external challenges (see above) they were all facing in the region. Utility managers were aware that regulations could change and utilities could benefit from working together to manage their shared challenges and provide a reliable water supply. Seattle, Tacoma, and Everett water utility managers began holding conference calls to discuss common challenges and soon these discussions were occurring monthly. When they decided to initiate action to create the Forum, they held a meeting with one of the mayors to enlist his support.

Following the identification and declaration of the benefits of collaboration, personal leadership and drive within the utility sector were major factors in overcoming obstacles and creating the Forum. Some of the obstacles faced by the Forum as it came together include state regulations, lack of authority by staff, utility resistance to loss of power, and funding.

There were no mechanisms in state laws that anticipated or facilitated creating an entity across three counties and involving a wide range of government agencies. One Act did allow agreements among local entities, and those pushing for the creation of the Forum stretched the conditions of the Act to create the local agreements that enabled creation of the Forum.

Because the Forum members are staff driven and are not elected officials, creating the Forum required (and still requires) significant coordination between staff and board members: board members have power to make and implement decisions, while staff members do not have any authority to act. Therefore, the process to make decisions is slow. However, because staff members are not elected, they have benefited by being able to speak candidly about difficult issues and suggest ideas for problem solving that elected officials might be reluctant to explore absent political backing.

Another obstacle for the Forum has been utility resistance to relinquish power. All members are accustomed to acting independently. As a group, decisions are difficult to make unless individual utilities broaden their perspective and think regionally instead of in their own best interest autonomously.

Funding was another problem in creating the Forum, however, the three large city water suppliers (Tacoma, Seattle, and Everett) committed to supplying the bulk of the funding. This settled the concerns of many of the smaller utilities considering membership.

Members of the Forum initially began meeting to improve communications, develop a consistent methodology for individual utility planning efforts, and where possible, to coordinate their individual planning efforts. These early objectives evolved to include efforts to address current and future regional public water needs and the water quantity needs of fish as they relate to public water supply. However, the Forum was not formed as a new regional governing body for public water supply. Rather, the Forum sets the context for collaboration to address issues of mutual interest. It does not have any specific authority over the provision of services: the member utilities remain separate entities. Within the Forum's membership is the recognition that there are a variety of water utilities in the region: cities, special purpose districts, water associations, private utilities, etc. These types of utilities are diverse in nature and they differ in specific interests and levels of autonomy.

The Forum is guided by the following objectives:

- To coordinate public water supply planning between water supply agencies

- To identify water supply needs defined by locally- and state-mandated Growth Management Plans
- To use water efficiently and seek new and better ways to conserve water
- To work collaboratively with local, state, tribal governments and stakeholders to ensure that water-based recovery portions of the ESA responses can be met in a timely and efficient manner
- To actively support a healthy fisheries program to be identified collaboratively by state, tribal and local governments as well as other groups
- To aggressively pursue environmental and other public benefits when developing public water supplies
- To actively pursue opportunities to implement these objectives on a regional basis
- To communicate with all groups that have an interest in water supply issues

One of the first accomplishments of the Forum is a first ever extensive assessment of water supply and demand in the three-county area. This assessment was documented in the Central Puget Sound Regional Water Supply Outlook (the Outlook), which was completed in 2001. It includes current and projected water demands (to 2020 for individual utilities, and to 2050 regionally), an analysis of regional water supplies, and details of potential future supply options. It also provides an assessment of projected municipal water supply shortfalls for individual utilities. Water suppliers, counties, and others will use this information to assist them in making water resource decisions.

Although commissioned by the Forum, the Outlook was possible only through a mutual effort that involved the surveyed utilities, the county governments, the region's watershed planning groups, and a variety of other stakeholders. In addition, the Forum solicited and received significant levels of technical support in preparation of The Outlook, including staff from two engineering firms and one independent consultant. Some of the technical support was paid, while most of it was volunteered. The technical support was instrumental in creating and conducting utility surveys, facilitating the work of subcommittees, preparing technical memoranda and assisting in the drafting of the final report.

Participants of the Forum are charged annual dues. Member agencies are able to leave the Forum at any time, at no cost (fees are no longer charged, upon departure), but as of 2005, no utilities of the Forum have ever opted to leave the group.

Constraints and Possible Solutions

Ensuring future water supply is one of the largest goals of the Forum. The Forum has identified some barriers to meeting municipal water needs. As utilities plan, select, and implement supply options, including water conservation and more effective management of existing resources, each supply option must meet considerable technical, financial, and environmental requirements. However, in addition to these basic water resource management requirements and reviews, there are institutional constraints and management issues that add difficulty to the review and approval process. In some cases, these constraints may preclude implementation of the most desirable utility supply options.

The Outlook defined an institutional constraint as "a rule, regulation, or policy, or the lack thereof, that unnecessarily results in additional cost, delay, or difficulty in implementing sound water resource management strategies and future water supply options." These constraints may encourage utilities to consider supply options that are easier to implement, but less economical and less environmentally responsible, with fewer regional benefits.

The Forum and its member water utilities, along with input from the Washington State Department of Health (DOH) and Department of Ecology, identified 12 significant institutional constraints faced by utilities, the state, local governments, and other stakeholders as the region develops and implements new supply options within a sustainable water resource management program. Some of the constraints are existing obstacles that impeded efficient decision making (e.g., lack of direction or information, uncertainty in planning), and the others are related to problems with the existing state Water Code that regulates the allocation of water (e.g., restrictions on moving water between utilities, difficulty in obtaining or changing water rights, validity and quantification of existing water rights). In the following sections, the constraints that have been assisted by the Forum's regional approach are discussed.

Balancing Available Information and Risk in Water Resource Decision Making

Water resource management decisions need to be made—potentially before adequate information can be available to make decisions with complete certainty about potential effects.

One possible solution is for the state to develop an adaptive decision-making process that allows water rights decisions to be made using the best available information and/or identifies and implements a program to obtain necessary information in a timely manner, on a schedule that balances risk with available information. Appropriate risk assessment and adaptive management would require that past decisions be changed if new information shows an unacceptable risk exists. If past decisions are changed through an adaptive management process, that process would also have to spell out how alternative solutions (supplies) are developed.

From a regional perspective, this issue has become the center of the Forum's work plan and members believe the enhanced collaboration and communication facilitated by the Forum is helping the decision making process. The Forum is directing their focus now on water rights related risks and on long term risks such as climate change (as the source waters supplying the Forum members depend on snow runoff to fill reservoirs).

Uncertainty of and Disincentives to Sharing Supplies

Some utilities have an available supply capacity that could be used on a temporary basis by other utilities until growth within the supplier's own service area requires that available supply capacity. However, utilities may be reluctant to sell water on a temporary basis to an adjacent purveyor if there is no guarantee in state law that ensures the supplying utility can stop supplying the adjacent purveyor at an agreed upon point in time. This does not apply to long-term wholesale agreements that are intended to continue in perpetuity.

This constraint represents a case in which the problem stems from a lack of code. A solution would be for the state to explicitly define legal "leasing" of blocks of water supplies in statute, where one utility could agree to supply another utility for a specific time period. This statute should include requirements for the utility receiving the water for the specific time period to have replacement supplies on-line by the end of the lease period and should specifically prohibit the use of a dependency argument to try to extend the duration of the supply beyond the terms of lease.

From a regional perspective, the Forum issues its newsletters to all elected officials and believes it collectively has a larger voice with which to influence regulatory change.

Limitations on Place of Use

Some utilities that could otherwise share supplies through inter-ties, which would defer or reduce the need for new supplies, previously could not do so because the place of use stated in their water right was too limited.

The Forum was instrumental in influencing a change in state law that enabled a place of use for a water right to be changed or updated through a utility's water system plan. This allowed flexibility in place of use, and allows utilities to "grow into" their existing water rights. The State of Washington historically issued a "pump and pipes" right which allowed utilities to acquire water rights if they showed the infrastructure was planned for use of water, even if they did not put the water to beneficial use yet. However, State laws cast doubts on the strength of this approach and utilities were concerned about their future rights. The Forum was instrumental in changes to the code which addressed this concern.

Other obstacles include autonomy and maintaining interest. While autonomy attracts membership, it also creates a challenge because utilities each have differing political agendas and sometimes it is difficult to align them. The Forum has also found that maintaining utility interest in the Forum is difficult absent a threat or major project. Water supply operations are running smoothly at the moment and utilities are showing less interest in the Forum. The Outlook was a good project bringing everyone together, but since then the Forum has had some difficulties maintaining participant interest. Utilities recognize the value in networking, but moving ahead on other less important Forum issues is difficult at times.

Conclusions

Several neighboring utilities facing common external challenges (e.g., existing and new regulations such as the ESA and GMA), instigated the formation of the Forum. Personal leadership and drive are credited with being the key contributing forces that made the Forum a reality. Most uncertainties with the GMA have been addressed by the Forum. However, some small systems may be unsure if they have sufficient long term supplies. ESA uncertainties (e.g., who is responsible for providing water, how much is to be provided) remain unresolved. The Forum is addressing the issue, but is not taking a pro-active role because there is a sense that water utilities should not lead this effort and should let the federal and/or state government and tribes lead the effort. However, many member utilities are comforted by having an organized regional body with which they can engage on this issue in a collective manner.

One of the factors leading to the lasting success of the Forum is that it was created to establish the context for collaboration to address issues of mutual interest, but not govern public water supplies regionally. It does not have any specific authority over the provision of services: the member utilities remain separate entities which are diverse in nature and differ in specific interests and levels of autonomy. This flexibility has proven to be a key factor contributing to the Forum's effectiveness in solving regional challenges. However, there are several institutional constraints faced by utilities, the state, local governments, and other stakeholders as the region develops and implements new supply options within a sustainable water resource management program. Nevertheless, through the pathways and structure for communication set up by the

Forum, the utilities have been able to collaborate to identify practical solutions and increase their influence collectively to make changes.

TAMPA BAY WATER, FLORIDA

Background

Tampa Bay Water (TBW) was created on October 1, 1998, as the new regional water authority for the Florida counties of Hillsborough, Pasco, and Pinellas; and the cities of New Port Richey, St. Petersburg, and Tampa. TBW supplies wholesale water to its member governments who in turn distribute it to their retail and wholesale customers through local water districts. It uses a mixture of groundwater, surface water, and desalinated seawater to provide 172 million gallons of water per day to more than 2 million people. The infrastructure consists of six groundwater treatment plants, 13 regional wellfields, a 66-mgd surface water treatment plant, a 15 billion gallon off-stream reservoir, a 25-mgd desalination plant, and over 200 miles of pipeline (TBW 2005). TBW is the sole and exclusive provider of water to it's member agencies (TBW 2005).

TBW is within the jurisdiction of the Southwest Florida Water Management District (SWFWMD). SWFWMD is one of five state regional entities responsible for maintaining an adequate supply of water and protecting the natural water supply systems. SWFWMD played a major role in the formation of TBW, and provides the authority with some capital funding. They have committed over $180 million to TBW's total investment, since 1998, of more than $750 million in new facilities. TBW also has received more than $57 million in federal commitments to date.

Motivation for Regional Solutions

TBW was created for the purpose of reducing groundwater pumping, developing new water supply sources, and obtaining funding from SWFWMD. The unsustainable, over-extraction of groundwater had caused severe infrastructure and environmental damage, and became the most pressing and visible issue motivating the formation of TBW as the regional water provider.

As early as the 1930s, the coastal regions of St. Petersburg and Pinellas County experienced salt water intrusion in their groundwater wells. Subsequently, the leaders of these communities bought wellfields in the more rural areas of northwest Hillsborough and central Pasco counties. During the 1970s and 1980s, Hillsborough and Pasco counties experienced a huge population growth and lake front communities were built within the cones of influence of the wellfields (Rand 2003). At the same time, the populations of St. Petersburg and Pinellas County also grew, resulting in increased water demand. In the early 1970s, Pinellas County conducted an aggressive land buying campaign in Pasco County for the development of wellfields. The groundwater pumping of the region grew 400% from 1960 to 1996, with over 20 billion gallons of water being exported from Pasco to Pinellas County (Dedekorkut 2003). This was the origin of what has been referred to as the "Tampa Bay Water Wars" (Dedekorkut 2003). In addition, water conflicts grew between St. Petersburg and both Pasco and Hillsborough counties.

Early attempts at ending drinking water disputes in the Tampa Bay area led to the formation of the West Coast Regional Water Supply Authority (WCRWSA), which was created by the Florida State legislature on October 25, 1974 (Dedekorkut 2003). It was created to acquire water and water rights, store and transport water, and deliver and sell water to its member governments

(Regan 2003). Over the next two and half decades, WCRWSA developed five major new water sources. However, it did not address the fundamental problem of reducing groundwater pumping, which continued as its sole type of water supply throughout its existence. Pinellas County, Tampa, and St. Petersburg continued to independently operate their water systems, including well fields retained by those governments (Dedekorkut 2003).

Over pumping of the Floridian aquifer caused claims of extensive damage to wildlife and the environment and destruction of land and property. Pumping led to lower water levels, and an array of related environmental damages, including soil oxidation and subsidence, increased invasion by exotic species, increased fire incidence, tree loss, and the loss of habitat for wetland dependent species (Regan 2003). SWFWMD determined that tens of thousands of acres of wetlands were damaged (Rand 2003). In a speech to SWFWMD, retired Green Beret Steve Monsees explained the destruction caused by the wellfields as follows: "…I am not complaining to you today of lowered lake water levels—but of the total and complete destruction of all water resources in our community…All surface water is gone. All wetlands and marshes are gone. Most wildlife has disappeared. The fish and the alligators are gone and now even the trees are dying…" (Rand 2003). Another distressed resident reported that her house had sunk 18 inches in 18 months (Rand 2003).

The groundwater disputes led to a $2 million litigation suit in the mid-1990s between the Tampa Bay municipalities and SWFWMD. SWFWMD offered the region $183 million for additional water supplies if they agreed to reorganize the regional water authority and end litigation (Rand 2003, Macpherson 2005). After having spent $2 million without obtaining any new water supplies, the municipalities chose to cooperate (Rand 2003) (the total cost was probably $10 million, including litigation and related costs). TBW replaced WCRWSA on May 1, 1998, when the local water agencies entered into the Amended and Restated Interlocal Agreement Reorganizing the WCRWSA (Raucher et al. 2005).

The formation of TBW was not voluntary at the start. SWFWMD exerted its authority to regulate groundwater pumping. The Florida State legislature also was involved. As described by Rand Baldwin, SWFWMD's Governmental Affairs coordinator:

> Basically it was a carrot-stick approach to the water supply. We basically said "OK, here's the deal. If you don't conserve and cut back on pumping groundwater out of the wellfields, we can fine you up to $10,000 a day. The carrot was We want to work with you and make $183 million available to you over 10 years to develop stable alternative water supplies—a desalination plant, reservoirs, interconnection of well." (Macpherson 2005)

What Happened: The Road to Success

Before 2002, the Tampa Bay region relied solely on groundwater from local well fields, with the sole exception of the City of Tampa, which extracted water from the Hillsborough River (Raucher et al. 2005, TBW 2005). After the formation of TBW surface and desalinated supplies were added to the dwindling groundwater reserves. The permitted groundwater pumping was reduced from 192 mgd to 158 mgd in 1998, and was further reduced to 121 mgd at the end of 2002. By 2007, only 90 mgd will be withdrawn from the wellfields, a reduction of over 53% (Raucher et al. 2005).

In order to comply with these groundwater pumping restrictions, TBW approved a series of master water plan projects including (TBW 2001):

- System Configuration I, providing 91 mgd to meet supply needs through 2007, at an estimated capital cost of $609 million:
  - 25 mgd seawater desalination plant on Tampa Bay (expandable to 35 mgd)
  - 60 mgd enhanced surface water system (15 billion gallon C.W. Bill Young Bay Regional Reservoir and 66 mgd capacity surface water treatment facility)
  - 6 mgd groundwater project
- System Configuration II and III, originally envisioned as providing 49 mgd to meet supply needs through 2014, at an estimated capital cost of $700 million (but modified at the time of this writing, January 2006, as indicated in parenthesis below):
  - 25 mgd seawater desalination on the Gulf of Mexico (Gulf Coast Desalination) (this has been shelved at the current time)
  - 5 mgd brackish water desalination (Mid-Pinellas Brackish Water Desalination) (this too has been shelved at the current time)
  - 10.5 mgd groundwater (Cone Ranch Wellfield—10 mgd, Eagles Well—0.5 mgd) (the Cone Ranch well field has been shelved at current time; but Eagles phase I was completed in 2004)
  - 3.5 mgd re-use of industrial process water (Crystals International Water Supply) (under active development)
  - 1.0 mgd reclaimed water exchange (Cargill Reclaimed Exchange) (under active development)
  - Hillsborough River augmentation during drought conditions (Morris Bridge Sink) (shelved at the current time)
  - 13.5 mgd Downstream Augmentation Project—use of reclaim water to augment the lower Hillsborough River and Tampa Bypass Canal to enable us to increase upstream freshwater withdrawals to supply the enhanced surface water system

Martin A. Rowland, author of "The Evolution of Two Water Resource Management Systems: Case Studies of Tampa Bay and the Middle East," credits TBW with providing the area with "the institutional means to acquire the additional water supply needed to meet its projected demand, while protecting the environment against adverse impacts, and operating within state and federal law" (Regan 2003, p. 162).

By obtaining substantial funding from both State and Federal sources, TBW has been able to acquire more expensive supplies of water while keeping its customers' prices relatively low. SWFWMD's co-funding will help lower the projected cost of desalinated water to customers (TBW 2005). As Honey Rand, President of the Environmental PR Group in Tampa Bay, explains, "With 54 inches of rainfall in an average year, what is there to fight about? In Florida it is not about available water, it is about cheap water" (Rand 2003).

Conclusions

The Tampa Bay region opted for a regional approach to water management for several reasons: (1) to reduce groundwater pumping and the extensive environmental and property damage associated with it; (2) to develop new water supply sources to reliably handle the continued population and economic growth of the area; and (3) to obtain the additional funding needed to allow for more expensive water supply projects. The district has had much success in reducing groundwater and increasing alternative supplies while keeping the cost to customers low.

Changing to a regional authority has had many benefits, but problems could still arise. As stated by Honey Rand, author of Water Wars: A Story of People, Politics, and Power, and professor of communications and environmental science at the University of S. Florida, "...the competition for cheap water will always be with us. At the same time, the means to exacerbate or resolve the resulting conflicts resides with us. Water resource allocation is a social undertaking. It is as much about the people involved as it is the resource" (Rand 2003).

CHAPTER 8
CONCLUSIONS AND A SUGGESTED AGENDA FOR FUTURE RESEARCH

This chapter provides a brief summary of key themes and findings (an abbreviated recap of the Executive Summary). Then, descriptions are offered of future research that would be most fruitful to the community of water suppliers, and those who work with them, as they face the challenges ahead and consider the role of regional approaches as potential solutions to those challenges.

SUMMARY OF KEY FINDINGS

As challenges faced for water utility sustainability increase over time, it is increasingly important that regional approaches be considered as potentially valuable ways to improve water service provision. There are a wide variety of options for regional approaches, ranging from simple cooperation by entities that remain completely independent, to two or more systems consolidating all assets and operations into a single regional entity, and including myriad options in between. This research has reviewed the broad array of options, and distilled them down to a simpler suite of six types of regional solution "agreements" that can be forged between water systems. These agreements become the focal point, as they serve as the mechanism through which the key elements of a regional approach are identified and addressed. In other words, the type and content of these agreements forge the basis for helping to ensure that the benefits of a regional approach can be realized, and that the issues or concerns associated with the approach also are recognized and managed in a suitable manner.

These six types of regional agreements are distinguished from each other along a gradient extending from sharing and collaboration at one end of the spectrum, to consolidated ownership and operations at the other extreme. The essential differences among the categories are anchored in what the participants are committing to do together. Our typology of regional solutions consists of the following:

- Mutual Aid Arrangements
- Sharing Arrangements
- Water Purchase Arrangements
- Collaborative Water Resource Development
- Contract Services Arrangements
- Consolidation

In addition, there are various lessons learned from the case studies about ways to approach and structure regional options in a way that can help promote successful collaborations. These elements that tend to promote regional approaches that succeed include (1) successful regional approaches are often driven by a clear recognition that a significant problem exists and that regional approaches can help solve it (and, the problem that predominantly leads to regional collaboration is a clear lack of adequate water supply); (2) regional approaches appear more palatable if each party retains ownership of certain key assets (e.g., local distribution systems and retail customers); (3) it is useful to have well articulated benefits that will be derived from a regional approach; (4) regional entities need to have sufficient authority to govern themselves and gain access to sufficient capital to finance investments in water and infrastructure that are necessary for sustainability; and (5) retaining some measure of local autonomy is a critical issue that frequently determines whether a regional approach will be considered favorably (and, regional approaches can be structured in ways that help ensure—rather than threaten—local autonomy).

AGENDA SUGGESTED FOR FUTURE RESEARCH

The research conducted for this project reveals several gaps that, if addressed by future efforts, are likely to help water agencies and other stakeholders as they consider options for regional approaches and strive to implement suitable arrangements.

1. **Empirical investigations of the benefits to be obtained from regional approaches.** One of the key obstacles identified to regional approaches is that customers, public officials, and other stakeholders tend to be skeptical about the benefits to be derived (e.g., cost savings and other advantages); they tend to believe the benefits will too small to justify making a change. Future research that provides a reasonably robust empirical assessment of the types and value of benefits, using a case study investigation, could (1) demonstrate that benefits can be demonstrated and be appreciable, and (2) provide an approach that can be replicated by utilities in other areas, to help them identify and assess the benefits of regional potential approaches. Case studies of failed agreements also would be beneficial from a learning perspective.
2. **Templates for the various forms of regional agreements that can be forged.** This research reveals the importance of developing suitable written agreements for any kind of regional collaboration approach. Utilities would benefit from having several "model" or template agreements that they can use as starting points for negotiating specific agreements within their region. As a starting point for developing such templates, research could be conducted to canvass water suppliers to obtain copies of existing agreements that have been found to be successful.
3. **Documenting the extent of existing water supply regionalization.** The water utility community may not fully recognize the level of existing supply regionalization or projected increase. A table of major utilities (e.g., those with over 1 million people served) and their wholesale customers (number, mgd, and percent of total production) would be informative and help utilities give greater consideration to regional solutions. It may be that virtually every very large utility provides service to at least one wholesale customer or has emergency inter-ties. This information can possibly be pulled together from AWWA's Water Stats database.

APPENDIX A
AN ANALYSIS OF REGIONALIZATION INITIATIVES FOR WATER MANAGEMENT: A CASE STUDY FROM THE PORTLAND, ORE., METROPOLITAN AREA

An Analysis of Regionalization Initiatives for Water Management: A Case Study From the Portland, Ore., Metropolitan Area

AwwaRF project 2950:
"Regional Solutions to Developing Water Supplies"

February, 2005

Prepared by
Natural Heritage Institute
Tellus Institute

A.1 INTRODUCTION

Background

Water utilities in the Portland, Ore., metropolitan area have worked together over the last decade to advance a regional approach to water purveyance and management as population growth and ecologic issues have placed greater stresses on available water resources. In this three county area, thirty-eight agencies provide water services to about 1.3 million people; the City of Portland's Water Bureau operates the water supply system that serves the majority (786,700) of these people, of which 482,500 are Portland municipal customers. Portland supplies water to the remaining 300,000 through wholesale contracts to 21 water providers (Figure A.1). The city's primary source of supply is the Bull Run Watershed, an area comprising 102 square miles of federally owned and protected land within the Mt. Hood National Forest and of which the City of Portland has retained sole water rights through state legislation for over a century. Water from this source is pure enough that no filtration is required to comply with current state and federal drinking water standards. The only treatment is the addition of chlorine and ammonia. Infrastructure in the watershed consists of two reservoirs with 10 BG of usable storage, a chlorine injection system, and the headworks to the conduit system that transmits the water to town (Figure A.2). Water flows by gravity to an in-town reservoir (50 MG) located at Powell Butte. The cities of Gresham and Rockwood receive Bull Run water from the conduits upstream of the Powell Butte Reservoir. At Powell Butte water can be added to the system from the Columbia South Shore Well Field (CSSW). The city relies on this source as a secondary and emergency supply. Up to 90 MGD can be derived from this source.

From Powell Butte, water is diverted into two major pipelines. One transmits water to the Mt. Tabor (61 MG) and Washington Park (34 MG) reservoirs for eventual delivery to city customers. The other pipeline delivers water to other parts of the city and communities in Washington County. In total, the system delivers over 38 billion gallons of water a year.

The Bull Run watershed typically experiences enough rainfall to keep the reservoirs full for 8 to 10 months of the year. It is only during summer that inflows are reduced and the capacity of the reservoirs is relied on to meet in-town demands. This also corresponds with the period in which demands are at their peak due to warm weather and landscape irrigation needs. For these reasons, most planning activities, with respect to infrastructure sizing and conservation programs, are focused on this "drawdown" period during which the reservoirs are not full.

Since 1996, the Portland Water Bureau and a number of other local water providers have explored cooperation through the "Regional Water Providers Consortium," an entity formed by intergovernmental agreement. The Consortium, comprising 23 water providers in the Portland metro-area, serves as a forum for cooperation on water supply planning, resource management, and conservation issues. In 2001, this dialogue evolved into a more concerted effort to create a formal regional water agency as many of the existing long-term contracts with the wholesale providers were approaching renewal. At this time, Eric Sten, the Portland City Commissioner with oversight responsibility for the Water Bureau, saw the opportunity to replace existing wholesale contractual relationships with a cooperative structure based on mutual ownership and stewardship of the area's water resources, including the Bull Run Watershed. The possible structure and composition of this regional water agency was debated and explored in two phases over an eighteen month period in 2001–2002, as the Water Bureau and thirteen other local water providers came together to document their needs and wishes for such an entity.

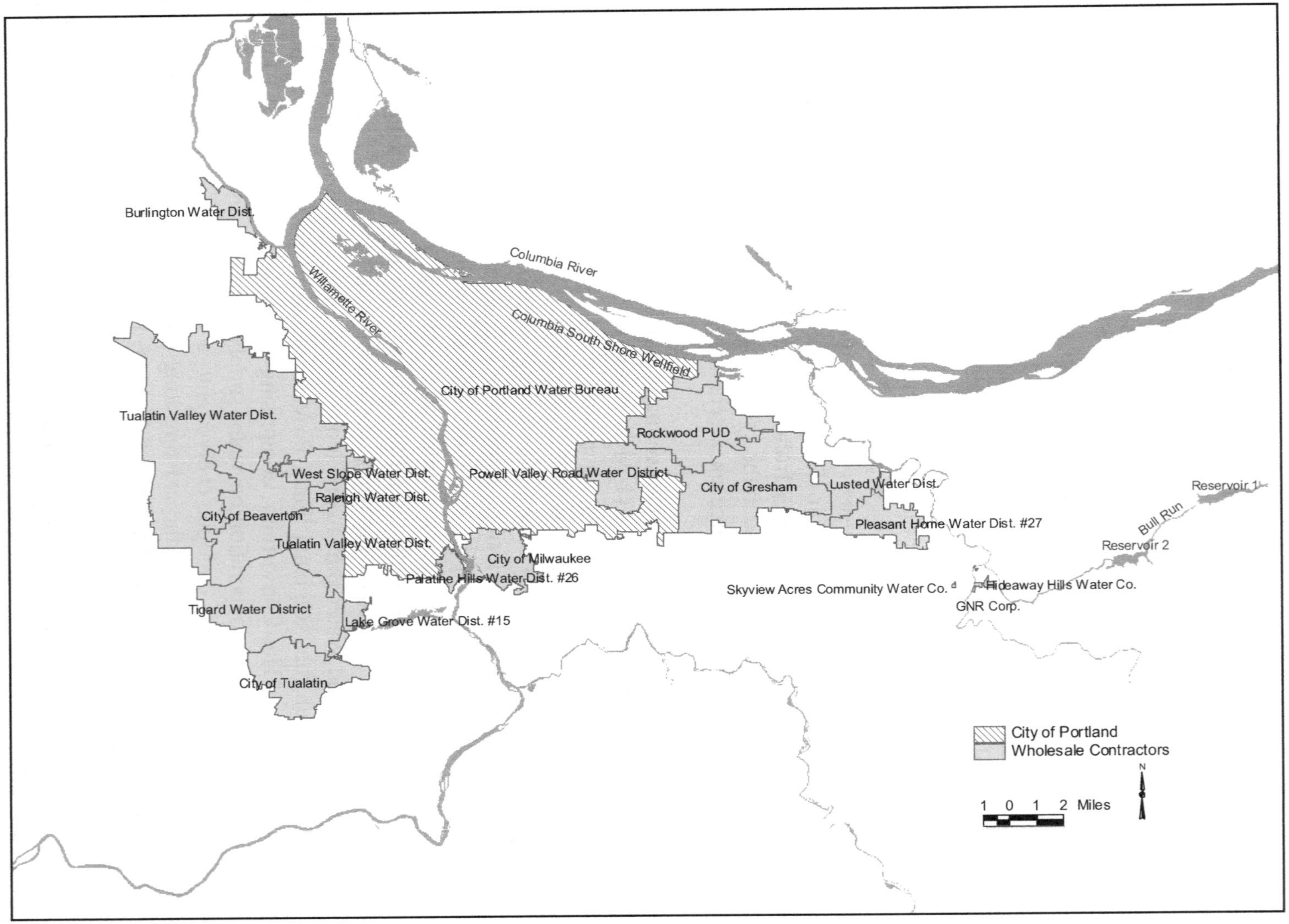

Figure A.1 Map of city of Portland and wholesale contractor service areas

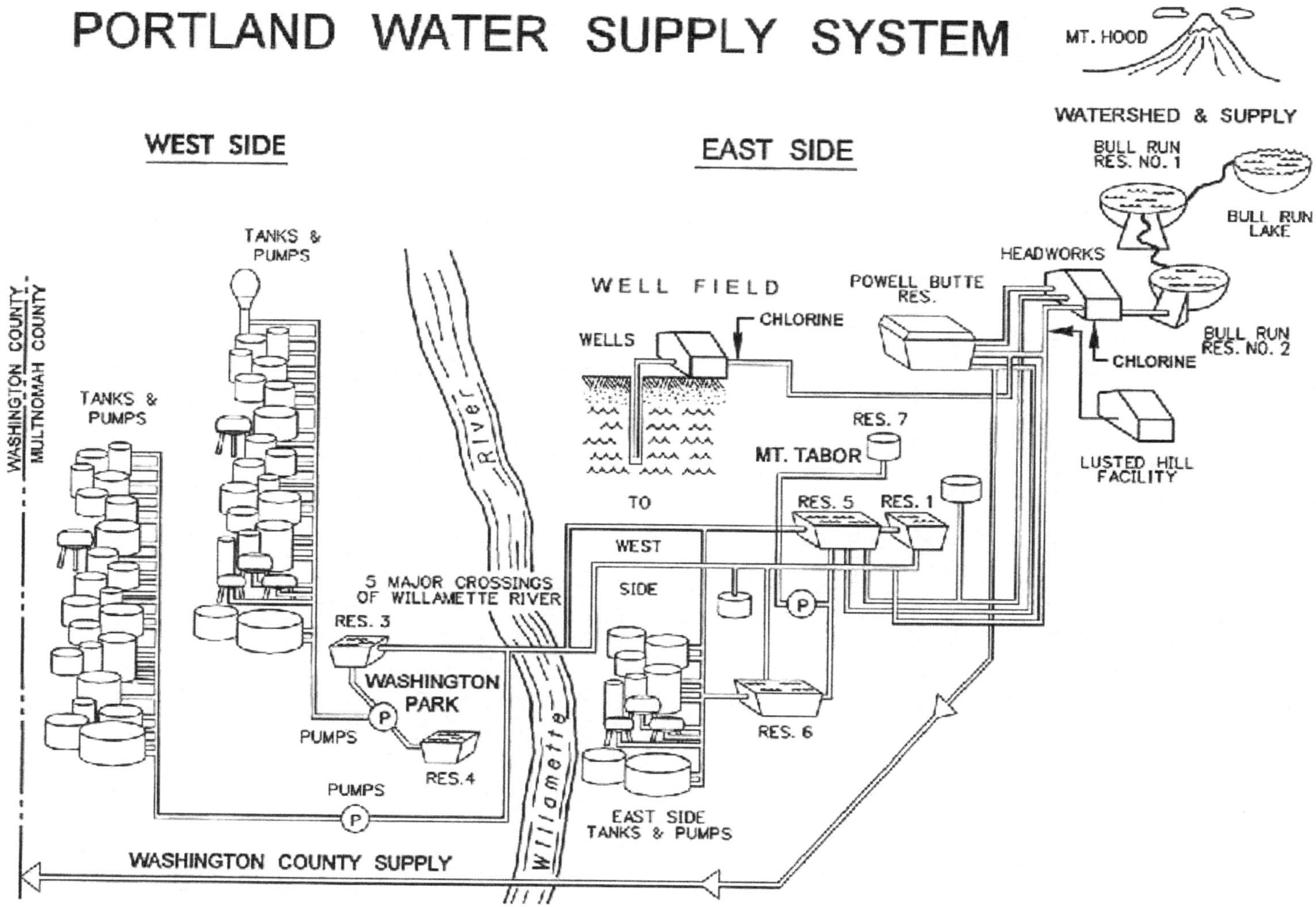

Source: Infrastructure Master Plan, City of Portland Bureau of Water Works 2001.

Figure A.2 Portland water supply system

The six-month long first phase of this exploratory effort began with several public meetings in which the regionalization effort was outlined and local water utilities' interest in creating a regional agency was gauged. This was followed by a series of weekly meetings in which senior managers from each of the participating utilities outlined possible governance models for the agency and synthesized common interests into a set of criteria that would serve as guidelines for formation of the regional agency. Phase I culminated in the identification of thirteen partner utilities, including the Portland Water Bureau, willing to commit to further pursuing the possible creation of this regional water agency.

During the second phase, which occurred during the first nine months of 2002, the thirteen partner utilities formed small "work groups" to analyze in greater detail the engineering, financial, and legal/governance issues that needed to be addressed and to amend the original set of organization criteria to reflect the input and concerns of local governments and private citizenry. From these analyses and input, a Principles of Agreement was drafted among the partnering utilities; this agreement was to form the basis of the new agency.

Unfortunately, this process of discovery did not culminate in the establishment of the envisioned regional agency. Within the span of several months, the initiative died due to a lack of resolution on several key issues identified during the two exploratory phases. These issues could not be addressed to the satisfaction of all the partnering utilities. In its place, talks began to renegotiate long-term contracts between the City of Portland and its wholesale customers.

Scope of Analyses

Below, we review the regionalization initiative that occurred in Portland during 2001–2002. In particular, we document the motivations for the initiative, elements of the process and issues revealed by the process that most likely led to the decision not to proceed toward implementation of the regionalization agreement, and lessons learned from this effort. These insights were gained through (1) study of documents obtained from the Portland Water Bureau which record the findings of the Phases I and II processes, and (2) communications with officials from the Bureau and a number of the partnering utilities and consultants involved in the regionalization effort. The bulk of these communications ensued during a meeting organized for this AwwaRF study in July, 2004, to which the representatives from the thirteen partnering utilities and the lead consultant were invited; these representatives included those who had contributed to both phases of the effort. While only five of the thirteen utilities and the lead consultant were represented at this meeting, we did obtain a broad range of insights on the factors that shaped the process and outcomes.

We also present an analysis of quantitative aspects of the regionalization process. In our discussions with partnering agencies, they expressed a belief that the process would have benefited from the use of a decision support system that would have allowed them to quantify and communicate particular supply and cost benefits that regionalization may have brought to the partnering utilities and Portland metro-area water customers. Here we outline the utility of such a decision support tool in this regionalization process, using WEAP (Water Evaluation and Planning system) to provide example demonstrations formulated from specific data acquired from the Portland metro-area water utilities.

In looking ahead to the future of water resource management in the Portland metropolitan area, we also relate a number of positive outcomes that stem from this aborted regionalization effort. Many of the partner utilities we talked with emphasized these outcomes as proof that, although the broad regional agency did not materialize, there will be lasting positive impacts for water management in the Portland metropolitan area that stem from this regionalization initiative.

A.2 REFLECTIONS ON THE REGIONALIZATION INITIATIVE

Motivations

Representatives from utilities that participated in the regionalization effort were asked to develop a list of the factors that most motivated them to engage in this initiative. Responses appear in Box A.1. A number of factors motivated utility participation in the regionalization effort; some motivations were shared universally among the partners, while other motivations originated from being on the wholesaler versus supplier (Portland Water Bureau) side of the regionalization effort. One of the most cited (six of eight meeting attendees) motivations for participating in the regionalization effort was the opportunity for shared ownership in a supply source. Many of the partnering utilities that were existing wholesale customers placed great value on gaining an ownership role and the accompanying ability to participate in decision-making on system improvements and priorities. From the Portland Water Bureau's perspective, the shared funding burden of capital improvements and expansions was also an important component to this ownership role.

After ownership, the most cited drivers for participation were the opportunity to conduct more efficient supply planning and the potential for cost savings through economies of scale (Portland noted that economy of scale savings may occur mostly through the water distribution operations). These savings would be manifest in lower long term rates for the metro-area customers. Portland also acknowledged that establishing a known customer base and supply through the regional agency could substantially enhance the efficiency of the planning process, as well as demonstrate the continued importance of the Bull Run system to the region's water needs, further ensuring its unique protected status into the future.

Box A.1
Motivations cited

- Develop an established customer base and supply
- Share cost of future system improvements
- Economies of scale (distribution, primarily)
- Cost savings
- Ownership role/responsibility for current wholesalers
- Expiration of current contracts
- Better, more efficient supply planning and system development
- Source diversity
- Supplemental sources
- Lower long-term rates for metro-area customers
- Opportunity for good governance/bolster public trust
- Reduce competition for water among municipalities
- Provide a participatory framework

Other motivating factors mentioned by the meeting attendees relate specifically to water supply or governance issues. These drivers include the opportunity to diversify water sources (in reference to reducing risks associated with shortages or interruptions that may strike one source or type of source), to provide supplemental sources, and to diminish competition among the utilities for water. Other attendees cited the regionalization effort as an opportunity to create a good governance structure and bolster public opinion in government. A related response mentioned the fostering of a participatory structure as a driving force.

Not all of the thirteen participating utilities were represented at the July, 2004 meeting, so documents recording the regionalization process were reviewed for additional motivating factors that may not have been cited by meeting attendees. We found that the statements of interest in the regionalization initiative that were submitted by the partnering agencies in the summer of 2001 to then Portland Commissioner Eric Sten broadly reflected the motivating factors voiced at the July, 2004, meeting and listed in Box A.2.

Box A.2
Regionalization initiative timeline, Phase I

Spring, 2001

- The idea of a regional entity that manages supply **and** distribution systems of Portland metro-area utilities is advanced by City of Portland Commissioner Eric Sten, who has oversight of Water Bureau; area utilities express interest, but most desire a structure that allows them to retain control of their distribution systems.

Summer, 2001

- Partnering agencies generate a list of criteria that will be used as guidelines for developing the regional agency.
- The partners advance a scope for the regional agency as one that will have the power to manage and control all pertinent regional water sources; the acquisition of Portland's Water Bureau and distribution system from the City of Portland continue to be included in this model.
- A preferred governance structure (ORS 190) is advanced; this structure allows for greater flexibility on many points but does not allow for direct election of board members.
- Public advocates reject the idea of including all regional water sources in the agency domain due to fears that without a direct voice in the governance of the agency, the public may be forced to use certain objectionable water sources, Willamette River water in particular, without their consent.
- City of Portland Commissioner Eric Sten suggests altering the model structure for the agency to include only management and control of the Bull Run watershed and Columbia South Shore well field.

Fall, 2001

- Fourteen area agencies agree to fund research for the formation of a regional drinking water supply agency
- Local, state, and national economic downturn is exacerbated by terrorist attacks on September 11. The economic climate, combined with ongoing chronic problems implementing a new billing system, produce financial concerns for the City of Portland and the Portland Water Bureau; public and City government confidence in the Bureau erodes.
- Responsibility for the Water Bureau is reassigned to City Commissioner Dan Saltzman, removing Commissioner Sten from leadership of the regionalization efforts.

Critical Factors Influencing the Outcome

We constructed an outline of the history of the regionalization effort after reviewing documentation of the process and interviewing Portland Water Bureau officials and representatives of the partnering utilities. This history is summarized in the timeline presented in Box A.2 (Phase I) and Box A.3 (Phase II) and lists the events and factors that most strongly influenced the outcome of the Portland metro-area regionalization initiative.

Box A.3
Regionalization initiative timeline, Phase II

Spring, 2002

- Thirteen partnering agencies enter into a funding agreement for the second phase of planning for the implementation of a regional water supply and transmission agency.
- Partners devise small work groups to provide details on engineering, financial, and governance issues related to forming the regional agency.

Summer-Fall, 2002

- Work groups generate a series of Technical Memoranda that provide legal, governance and financial details to be considered with formation of new agency.
- Because most assets to be assigned to the agency are owned by the City of Portland, a "buy-in" strategy is devised in which partners will be required to proportionally reimburse the City of Portland for these assets based on their intended use of the Bull Run/well field supplies.
- An initial valuation of the assets that will be assigned to the new regional agency is produced based on simple accounting methods; this procedure obtains a valuation of approximately $180,000,000.
- Portland Water Bureau officials conduct a more detailed valuation of the Bull Run/well field assets, refiguring based on more realistic depreciation life estimates; this method obtains a valuation of approximately $320,000,000. Water Bureau officials recommend that an external valuation be conducted to provide a more thorough, due diligence valuation.
- A Principles of Agreement is constructed, guided by initial criteria developed in Phase I and informed by the information gathered in Phase II by the work groups.
- City of Portland Council and Mayor are presented with the agency plan, along with the valuation of the Bull Run and well field assets. Sobered by the economic hardships of the previous year, the City government expresses concern over the substantial loss of the Bureau and its long-term revenue-generating capacity for the City and fear that the valuation of the Portland assets will be insufficient to reimburse the City when considered in this long-term context. They also question the structural balance of the agreement, questioning why Portland is the only participant contributing its assets/distribution system to the new agency.

Winter, 2002–2003

- City of Portland Council and Mayor withdraw their support of the regionalization initiative due to financial considerations.
- City of Portland Mayor formally invites wholesales to participate in a new negotiation process to develop long-term contracts between the Portland Water Bureau and its wholesale customers.

Box A.4
Critical factors determining outcome

- Reimbursement (buy-in) cost for Portland assets too high
- CoP reluctant to relinquish revenue-generating Water Bureau to agency
- Key CoP staff (e.g., financial) not involved in process
- Initial political champion of process did not remain in elected position
- Public pressure opposed preferred agency structure
- CoP government not engaged as actively in process
- Some key decision-makers did not share goals of the partnering utilities
- Economic climate, financial pressures on CoP and Portland Water Bureau at time
- Lack of public confidence and advocacy for effort
- No strong drivers (economic, regulatory, crises) for effort

Several interlinked factors appear to have ultimately prevented the implementation of the Principles of Agreement developed by the partner utilities. As discussed below, these factors relate largely to the evolution of the proposed agency scope; this scope eventually included only the Bull Run and Columbia South Shore well field supply system assets owned by the Portland Water Bureau. The initiative ultimately died trying to address the question of how, or if, the City of Portland could be adequately compensated for the transfer of these assets and the loss of the revenue potential of the Water Bureau, which would be transferred with these assets to the new agency.

Responding to an inquiry as to why the initiative ultimately failed to create the regional agency, the partnering utilities that attended the July, 2004, meeting provided their own list of critical factors. These factors are listed in Box A.4. Most attendees cited (five of the eight) the buy-in cost for the partners as too high. In other words, the compensation to be paid to the City of Portland for the value of the Bull Run, Columbia well field, and transmission assets to be assigned to the regional agency was higher than they believed it should be. Many of the partnering utilities suggested they had already substantially compensated Portland for these assets through depreciation payments to the City made under their wholesale contracts over the prior three decades.

The reluctance of the City of Portland to relinquish ownership of the Water Bureau and the Bull Run and well field assets because of their long-term revenue generating potential was also cited by five attendees as a critical factor in ending the effort. Many believed the City had not fully comprehended this revenue factor when they formally supported the regionalization effort at the outset of Phase I; the economic downturn of the last two years had underscored the City's reliance on Water Bureau revenue that they believed could not be adequately compensated for by a regionalization "buy-in" amount that would be acceptable to all parties.

Several attendees believed that a strategic error had been made in not including key City of Portland stakeholders, particularly the Office of Financial Management, in the Phase I and II discussions, where the valuation question could have been handled earlier and with broader input from the City of Portland administration. This strategy had been put forward by the early champions of the effort, to minimize risks of the initiative getting bogged down by details and losing momentum. However, as the proposal moved forward into Phase II, the City's financial administration and the public began to view the initiative with concern. A lack of confidence in the Water Bureau that had developed during an ongoing billing system crisis grew to include the perception that the Bureau could not negotiate the regionalization process with the best interests of the City in mind. This lack of confidence in the Water Bureau was highlighted in late 2001 when the Mayor reassigned responsibility for the Water Bureau to Commissioner Dan Saltzman. This reassignment removed the original champion of the regionalization effort, Eric Sten, from a critical leadership role in the regionalization efforts at a critical point in the process.

The issues surrounding the valuation of the Bull Run assets were exacerbated by the City government's perception of the negotiated agency scope, which had evolved from that of a broad entity incorporating all pertinent water sources and distribution systems belonging to the partnering utilities to one comprising only City of Portland assets. The City's government and Portland citizenry viewed this as an unbalanced agreement, with the smaller partnering utilities appearing to contribute little to the new entity while the City of Portland was giving up control of a substantial revenue-generating utility.

Several key factors had driven the evolution of the proposed scope for the agency that was ultimately submitted to the Portland City Council and Mayor for consideration. Although the initial proposal set forth by Commissioner Sten in March, 2001, suggested formation of an agency with broad control over regional supplies, transmission and distribution, partnering utilities indicated their desire early to retain control of their distribution and billing structures. From their perspective, the scope of the agency should center on regional water *supply and transmission* only. Thus, the inclusion of individual utilities' distribution systems within the control of the agency was left as an *option* for the future. The distribution system owned by the Water Bureau continued to be considered for probable transfer, along with the Bureau, to the new agency.

All partnering utilities saw substantial long-range benefit to bringing control of the various water sources under one regional agency. However, early public pressure originating from the fear of potentially receiving water from undesirable local sources, specifically the Willamette River, encouraged the agency planners to limit even the supply scope of the agency. Public detractors of the agreement highlighted the proposed governing structure of the new agency, which did not include direct election of agency board members, in making their argument for a limited supply scope for the agency. As a result, the guideline criteria for the new agency were amended by the end of Phase I so that the supply scope would comprise only the Bull Run and Columbia well field supplies—both owned solely by the Water Bureau/City of Portland. In the end, the City of Portland administration declined to accept the proposed agreement due to the perceived imbalance in agency scope and the fear of losing the revenue-generating power of the Water Bureau without adequate compensation. In turn, the city proposed negotiation of new long-term contracts with wholesale customers in lieu of further discussions centered on formation of a regional agency.

Lessons Learned

Partnering utilities were asked to reflect on the regionalization initiative with respect to what important lessons they felt were learned by conducting the process. Many also took this opportunity to contribute suggestions for different approaches that may have led to a successful outcome. These lessons and suggestions are outlined in Box A.5.

A number of utility representatives believe that if given another regionalization opportunity, they would engage critical stakeholders much earlier, particularly the staff of the Portland Office of Financial Management. Portland financial management representatives were asked to review the asset valuation results toward the end of Phase II of the initiative, instead of being asked to contribute to this process early on. As discussed above, this was the result of a strategic

Box A.5
Important lessons

- Engage critical stakeholders at outset (e.g., CoP financial)
- Having a motivation external to the utilities would help
- Public in Portland metro-area doesn't appear to be behind a regionalization effort at present
- Need to communicate a clear motivation and public benefit
- Develop proper context for public engagement in process
- Anticipate that it takes a long time to plan for the details
- Need to adapt to changing political climate and conditions
  –Work teams were essential for providing needed data
- Commitment to working together to "see it through" is critical

decision to maintain momentum for the negotiations. However, the strategy coupled with other factors, including the billing system failure and downturn in the local economy, ended up producing a lack of confidence by the City of Portland government both in the Water Bureau and the regionalization proposal in general. Key officials apparently were "surprised" late in Phase II when the proposed agency structure and financial details were fully conveyed to them.

Representatives also indicated that having some sort of strong external driver for the regionalization, such as a regional water shortage, would have provided a driving force to push such an initiative to implementation. Simply put, the Portland metro-area public didn't perceive any compelling need to pursue regionalization. In order to get all stakeholders—city government, the public, local commerce and utilities—behind such an effort, either some sort of crisis (e.g., a water shortage or threat to the supplies) or active promotion by other stakeholder groups was needed.

Many representatives stressed the lesson that strong driving forces and motivations have to be clearly communicated to the public and other stakeholders early on to build political and public momentum behind the initiative: "If it is clearly a good idea, the stakeholders will rally." However, some of the group also indicated they would have preferred to develop some elements of the plan more completely before presenting it for public consideration. Some believed that having the early discussions open to the public hampered the process because at that point the utilities themselves did not really know what form this regional entity should take. They believed the public viewed these early discussions as demonstrating indecisiveness and a lack of clear direction and vision. These perceptions bred a lack of confidence in the process among the public.

The representatives also now realize that such an effort takes much longer to implement than they initially expected. They now see great future value in the engineering and financial data they acquired for and from the process. They admit, though, that they substantially underestimated what would be needed to plan such an undertaking. The partners believe that if they had known the length of time that would be required to pull together an agreement, they would have more effectively planned for navigating the process through the changing political and economic conditions that occurred and which had a substantial impact on their bargaining position with the City of Portland.

Finally, the partnering utilities wanted to stress the importance of the revenue factor in the eventual decision to not continue efforts that may have led to formation of a Portland-area regional water agency. The City of Portland reacted very negatively to the potential loss of future revenues from transfer of the Water Bureau to the proposed regional agency. Because many city-owned water utilities provide substantial general revenue to the municipalities they serve, this factor could play an important role in other regionalization efforts across the country. From a policy and political perspective, foreseeing and addressing this key financial consideration could aid in ensuring a regionalization effort's success.

A.3 QUANTITATIVE ASSESSMENT

Introduction

In this section we explore the utility a decision support system might have provided to the regionalization process had it been available. The analysis presented here is by no means a complete analysis of the possible demand or supply configurations that could be utilized in the Portland metropolitan area. Instead, its purpose is to illustrate the types of information that would have been available if a decision support system had been used.

The analysis was undertaken from the perspective of communities neighboring Portland that were considering the costs and benefits of joining the regional system. While the costs of buying into the Portland system were becoming clear through the determination of the assessed value of Portland's assets and preliminary estimates of the cost of potential upgrades, less information was available on the reliability of the upgraded system under extremely dry conditions. The communities analyzed here had the option of relying on the Portland system for all or a portion of their supply needs and, in fact, had wholesale contracts with Portland at the time of the regionalization discussions. If Portland were to provide only a portion of the supply to these communities, sources other than the Bull Run system and the Columbia South Shore Well Field would need to be developed. The example problem explored here shows how a decision support system could provide an estimate of the reliability of Bull Run Supply and the Columbia South Shore Well Field and the costs of improving that reliability. The contractors could then use this information to inform decisions regarding their level of involvement in the regional entity and other supply development. The analysis described here was an exploration of some of the following issues for the wholesale contractors of Tigard, Beaverton, Tualatin, and the Tualatin Valley Water District (TVWD) (Figure A.1):

1. The types of infrastructure required to satisfy projected demands; possibilities included expansion of Bull Run storage by raising Dams 1 and/or 2, expansion of in-town storage at Powell Butte, and expansion of pipeline capacity.
2. The reliability of the Portland water supply system in meeting the demands of the contractors assuming all City of Portland demands were met first.
3. The projected costs of infrastructure enhancements necessary to improve the Portland water supply system and increase system reliability.
4. The ramifications of altering the levels of supply provided to the contractors by the Portland water supply system.

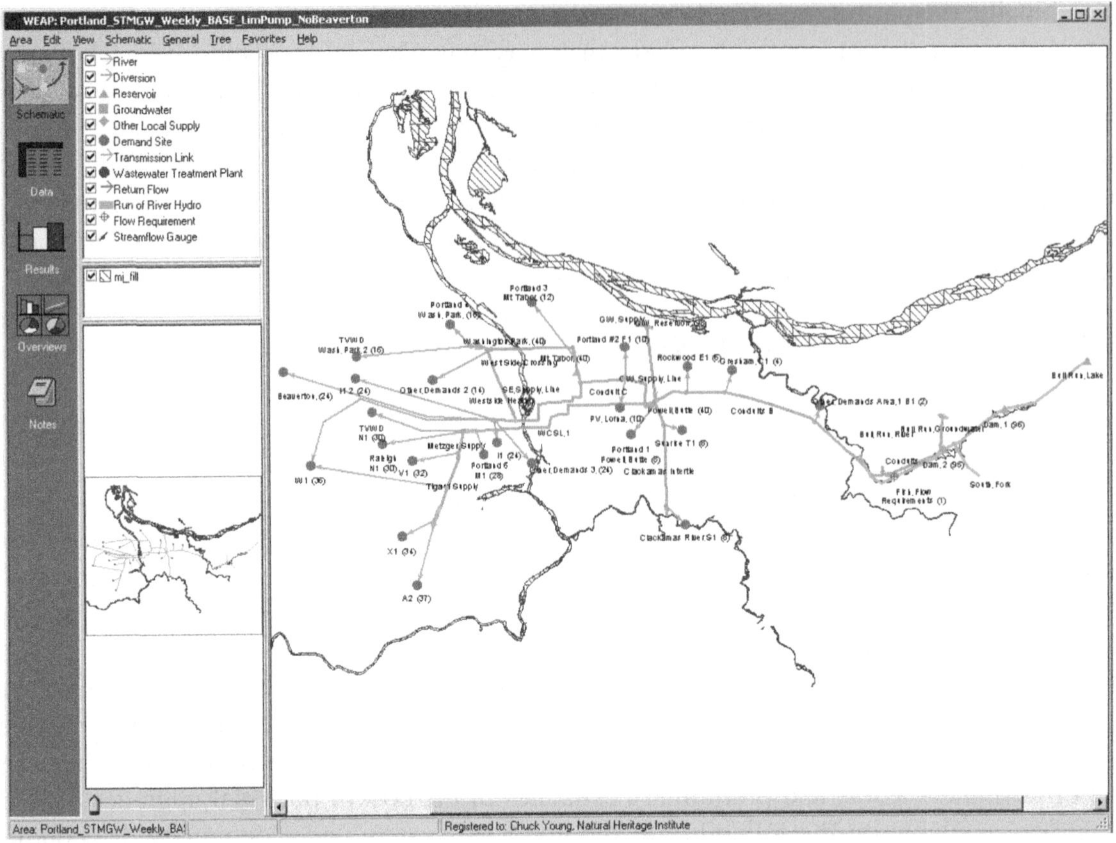

Figure A.3 WEAP model schematic

Model Description

A model of the Portland system was developed in the Water Evaluation and Planning System (WEAP). WEAP is a water balance tool developed by the Tellus Institute and the Stockholm Environment Institute (http://www.weap21.org/). A WEAP model (Figure A.3) was developed that included the Bull Run supply system, demand nodes for the City of Portland, and demand nodes for communities with wholesale contracts (see Attachment A.1).

The WEAP model allows the user to set the priority by which demands will be met. Required minimum flows on the lower Bull Run were modeled using the proposed rules provided by the Bureau (Attachment A.2) and given the highest priority. This effectively meant that during periods of supply shortages some demand nodes could have unmet demand before the flow requirement is violated. For demand nodes, priorities were set so that nodes closer to the Bull Run system received all needed supply before demands further from the Bull Run system received supply. Effectively, this meant that upstream demand nodes were less likely to experience shortages than downstream nodes.

Assumptions

- Pumping from the CSSW was initiated once the seasonal drawdown of the Bull Run reservoirs began. Initially pumping was set at 35 MGD. This rate was maintained for the first month. Following that, and until the reservoirs began to refill, the pumping rate was 90 MGD. If the total volume pumped in a year exceeded 12 billion gallons, pumping was halted. If the 12 BG limit was not reached pumping was halted once the reservoirs began to refill.
- Model scenarios were run from 2005–2025.
- Demand data were obtained from the Bureau and reflect current projections of population increases.
- Bull Run system inflow data for 1987 were used which, according to the Bureau's Infrastructure Master Plan, has a historical recurrence interval of 1 in 30 years when peak season impacts are considered. 1987 was chosen since it represents an extreme case with respect to summer inflows and would be a good test of system reliability.
- The model was run using a weekly time-step. The availability of daily data did make a daily model possible, however, a weekly time-step was used to keep the simulation run-times manageable. The possible negative consequence of using a weekly time-step, in comparison to a daily time-step, is that the smoothing effect introduced by weekly averaged values could result in an under-reporting of pipeline capacity or supply constraints since these can occur at shorter time-scales than a week.
- All demands from Tigard and Tualatin were served by the Portland system.
- 50% of Tualatin Valley Water District demands were served by the Portland system.
- In scenarios including Beaverton, 2 MGD of demand were served by the Portland system.
- All increases in Dam 2 storage (2.2 BG) were allocated to meeting municipal demands.
- The Tigard demand node (W1) was given the ability to receive supplies from the Tigard Supply Line via the Washington County Supply Line as a first preference and from the Southeast Supply Line as a second preference. This reflects the actual system, as water can be delivered to Tigard via gravity through the preferred route. If a capacity constraint is reached on the Washington County Supply Line or the Tigard Supply Line, demand can be met from the Southeast Supply Line, however, pumping is required.

Water Supply Reliability Scenarios

Scenario 1

The first run used the demand and infrastructure configuration detailed in Section 3.3. This was done to determine when and where shortages would occur during the 2005–2025 time period assuming all minimum flow requirements were met (Table A.1).

Unmet demands first occurred during the year 2020 for Tigard, TVWD and Tualatin. Beaverton had unmet demands starting in 2021. Beaverton had zero unmet demand in 2020 due to the fact that it is closer to the sources of supply and that delivery priorities decrease in the downstream direction. No shortages were caused by flows reaching pipeline capacities, however, the

Table A.1
Unmet demands for wholesale contractors caused by shortage of supply (MG)

Year	Beaverton	Tigard	TVWD	Tualatin
2005	0.00	0.00	0.00	0.00
2006	0.00	0.00	0.00	0.00
2007	0.00	0.00	0.00	0.00
2008	0.00	0.00	0.00	0.00
2009	0.00	0.00	0.00	0.00
2010	0.00	0.00	0.00	0.00
2011	0.00	0.00	0.00	0.00
2012	0.00	0.00	0.00	0.00
2013	0.00	0.00	0.00	0.00
2014	0.00	0.00	0.00	0.00
2015	0.00	0.00	0.00	0.00
2016	0.00	0.00	0.00	0.00
2017	0.00	0.00	0.00	0.00
2018	0.00	0.00	0.00	0.00
2019	0.00	0.00	0.00	0.00
2020	0.00	46.59	32.04	40.13
2021	14.00	65.91	102.71	41.49
2022	14.00	203.48	259.77	145.75
2023	15.95	266.30	441.47	257.95
2024	25.06	336.38	631.15	336.92
2025	42.35	408.92	791.72	423.91

pipeline that serves Tualatin and Tigard (20 MGD max.) reached maximum flow for part of each year starting in 2006. Since Tigard demand could be met via the Southeast Supply Line during these periods, no shortages occurred. The economic tradeoffs of pumping via this route versus expanding the gravity fed pipeline were not explored in this study.

Since all of the unmet demands were caused by supply shortages the next model run was formulated with an increase in storage on the Bull Run system by raising the crest of Dam 2 from 860 to 872 ft. This increase in height resulted in a 2.2 billion gallon increase in storage. The model was run with the Dam 2 storage increase brought on-line in 2020 (Table A.2).

With the increase in Dam 2 storage, unmet demands did not occur until 2025. Beaverton had no shortages for the entire simulation. The other three contractors had shortages in 2025 that were an order of magnitude smaller than the shortages calculated without the increase in Dam 2 storage.

Table A.2
Unmet demands with Dam 2 raised in 2020 (MG)

Year	Beaverton	Tigard	TVWD	Tualatin
2005	0.00	0.00	0.00	0.00
2006	0.00	0.00	0.00	0.00
2007	0.00	0.00	0.00	0.00
2008	0.00	0.00	0.00	0.00
2009	0.00	0.00	0.00	0.00
2010	0.00	0.00	0.00	0.00
2011	0.00	0.00	0.00	0.00
2012	0.00	0.00	0.00	0.00
2013	0.00	0.00	0.00	0.00
2014	0.00	0.00	0.00	0.00
2015	0.00	0.00	0.00	0.00
2016	0.00	0.00	0.00	0.00
2017	0.00	0.00	0.00	0.00
2018	0.00	0.00	0.00	0.00
2019	0.00	0.00	0.00	0.00
2020	0.00	0.00	0.00	0.00
2021	0.00	0.00	0.00	0.00
2022	0.00	0.00	0.00	0.00
2023	0.00	0.00	0.00	0.00
2024	0.00	0.00	0.00	0.00
2025	0.00	48.43	69.00	47.02

Scenario 2

In an effort to increase system reliability to 100% over the 2005 to 2025 time period using 1987 conditions, a scenario was developed in which it was assumed that the City of Beaverton did not participate in the regionalization effort and obtained its water from other sources. Bureau staff indicated that this was a possible since Beaverton has access to other sources of supply.

The model was run with the same parameters as in Scenario 1, current infrastructure, and the removal of the 2 MGD Beaverton demand. Results from this model run are in Table A.3.

Unmet demands started in 2021 in contrast to the model run that included Beaverton and had shortages starting in 2020. No unmet demands resulted from pipeline capacity constraints, however, each year starting in 2006 Tigard did receive some water from the Southeast Supply line when the pipeline serving Tigard and Tualatin flowed at maximum capacity.

To see the effect of the Dam 2 expansion on reliability, another model run was conducted with the Dam 2 expansion coming on-line in 2021. There were no unmet demands during the

Table A.3
Unmet demands with Beaverton removed from system (MG)

Year	Tigard	TVWD	Tualatin
2005	0.00	0.00	0.00
2006	0.00	0.00	0.00
2007	0.00	0.00	0.00
2008	0.00	0.00	0.00
2009	0.00	0.00	0.00
2010	0.00	0.00	0.00
2011	0.00	0.00	0.00
2012	0.00	0.00	0.00
2013	0.00	0.00	0.00
2014	0.00	0.00	0.00
2015	0.00	0.00	0.00
2016	0.00	0.00	0.00
2017	0.00	0.00	0.00
2018	0.00	0.00	0.00
2019	0.00	0.00	0.00
2020	0.00	0.00	0.00
2021	46.98	102.71	41.49
2022	95.32	142.81	87.51
2023	205.06	312.48	196.45
2024	267.93	523.64	265.61
2025	339.52	687.04	348.18

entire simulation. This is in contrast to Scenario 1 in which the model run with expanded Dam 2 had shortages during 2025.

Financial Analysis

An analysis of the costs associated with infrastructure development was conducted to illustrate how the use of a water balance tool such as WEAP can be used inform a financial analysis of supply options. The objective of the analysis was to illustrate the financial impacts to contractors of upgrading the system so that delivery reliability was 100% during the 2005–2025 time period using 1987 inflows. Capital and O&M costs were provided by Bureau staff. The distribution of these costs among the contractors of interest was done using cost distribution factors derived from Exhibit 7 of a memorandum by the Financial Consulting Solutions Group Inc., dated December 2, 2003, to Bernice Bagnall of the Tualatin Valley Water District. The distribution factors (Table A.4) were calculated as the ratio of contractor specific costs to total project

Table A.4
Cost of infrastructure development (2000 dollars) and cost distribution factors

	Total	TVWD	Tigard	Tualatin	Beaverton	Others
Washington County Supply Line Expansion	$112,721,022	$30,998,281	$8,115,914	$9,581,287	$3,607,073	$60,418,468
Powell Butte Expansion	$46,950,290	$12,911,330	$3,380,421	$3,990,775	$1,502,409	$25,165,355
Conduit 5	$18,735,357	$5,152,223	$1,348,946	$1,592,505	$599,531	$10,042,151
Raise Dam 1	$1,000,000	$275,000	$72,000	$85,000	$32,000	$536,000
Raise Dam 2	$10,000,000	$2,750,000	$720,000	$850,000	$320,000	$5,360,000
Cost distribution factor		0.275	0.072	0.085	0.032	0.536

costs. The costs shown in Table A.4 for the Washington County Supply Line, Powell Butte expansion, and Conduit 5 were derived from the memo. In the memo, costs were presented in 2012 dollars. These values were discounted to 2000 dollars assuming a 3% discount rate. The values in Table A.4 for raising Dams 1 and 2 were provided by Bureau staff and are in 2000 dollars. The "Others" column in Table A.4 represents the costs to the remaining users of the Portland supply system, a vast majority of which, on a volume basis, are City of Portland customers.

Scenario 1 Financial Analysis

To calculate the cost of raising Dam 2 and bringing it on-line in 2020, estimates of the capital cost and operation and maintenance (O&M) costs from 2000 were adjusted to 2020 dollars using an inflation rate of 2.5%. The original (2000) construction cost estimate was for $10,000,000 and O&M was estimated to cost $50,000 per year. In 2020 dollars, the construction cost was $16,386,164 and annual O&M was $81,930. It was assumed that capital costs were financed for 20 years at an interest rate of 5.5 % compounded annually. O&M costs were assumed to be constant for the 20 years analyzed (2020–2039). The total annual cost of interest payments, principal repayment, and O&M expenditures was $1,453,114. These annual payments were discounted back to 2005 dollars using a discount rate of 5.5% and summed for a total project cost of $8,206,294. Using the cost distribution factors shown in Table A.4, the costs of raising Dam 2 (2005 dollars) for each of the wholesale contractors was calculated (Table A.5).

Scenario 2 Financial Analysis

The cost of raising Dam 2 in 2021 was calculated using the approach described above for Scenario 1. The adjusted cost for construction in 2021 was $16,795,818. Annual O&M costs ($50,000 in 2000) adjusted to $83,979 in 2021. Total annual costs for construction and O&M

Table A.5
Cost shares for raising Dam 2 (2005 dollars)

	TVWD	Tigard	Tualatin	Beaverton	Others
Scenario 1—raise Dam 2 in 2020	$2,255,089	$594,243	$699,176	$262,601	$4,395,183
Scenario 2—raise Dam 2 in 2021	$2,263,392	$596,431	$701,750	—	$4,411,366

summed to $1,489,441. These annual costs, for years 2021–2040, were discounted to 2005 dollars for a total cost of $7,972,939. The distributed costs are shown in Table A.5.

The financial results from Scenarios 1 & 2 show that the removal of Beaverton from the system resulted in a 0.3% increase in costs for the remaining contractors. This effect was somewhat reduced by the one-year delay of Dam 2 expansion. The additional cost improved reliability so that there were no shortages in 2025.

Conclusions

A decision support system such as WEAP can serve a useful purpose in the negotiations for the creation of a regional water supply entity. Such a tool can be used to study the projected impacts to various stakeholders in the negotiation process. The analysis described here illustrated how WEAP could help stakeholders quantify the reliability of the water supply system both in terms of pipeline capacities and storage volumes. Had it been used, WEAP would have allowed the stakeholders to explore various configurations and timing of supply infrastructure and demand levels and to calculate supply reliability and financial impacts.

A.4. LOOKING AHEAD

The regionalization initiative advanced by the Portland Water Bureau and thirteen metropolitan area water suppliers during 2001–2002 grew out of cooperative spirit and the desire to implement a more efficient, cost-effective planning framework around water. Building on relationships nurtured over a number of years through long-term wholesale water contractual arrangements between the City of Portland and metro-area water suppliers, this regionalization effort was an attempt to make more concrete the coordinative mechanisms first established between the parties with the Regional Water Providers Consortium in 1996. With many long-term contracts set to expire in the near future, all parties felt the time was right to consider establishing a formal entity to jointly manage the region's water resources. Motivated largely by the desire for shared ownership and decision-making responsibility for the Bull Run watershed, the region's primary water supply, the PWB and thirteen of its wholesale contracting partners worked together for over 18 months to research the operational, financial, and governmental implications for this shared responsibility.

Although a lack of resolution on several key issues prevented moving forward to implement the formal regional agency framed during this discovery process, both the PWB and its wholesale contracting partners believe that future water supply planning in the region will benefit greatly from other products of this regionalization effort. A primary example is the infrastructure and financial data collected during this effort, which is proving valuable in informing the renegotiation of long term contractual agreements between the PWB and the wholesale partners. Both the PWB and its wholesale partners also tend to agree that the relationships among metro-area water providers were strengthened by working together toward the common goal of determining what a regional water agency should "look" like for the Portland metro area. The process provided a forum for sharing perspectives, ultimately fostering an enhanced level of mutual understanding among the water suppliers that will serve them well as they face important decisions regarding the future allocation, distribution, and protection of the region's water resources.

A.5 ACKNOWLEDGMENTS

We would like to thank the many Portland metro-area water utility representatives who contributed their ideas and thoughts to this review of the regionalization initiative that took place there in 2001–2002. Those representatives that attended our fact-finding meeting in July, 2004 included Jerry Arnold (West Slope Water District), Greg DiLoreto and Todd Heidricken (Tualatin Valley Water District), Mark Knudson (Portland Water Bureau), David Rouse and Dale Jutila (City of Gresham), David Winship (City of Beaverton), Ruth Spetter (City of Portland City Attorneys Office), and Phil Smith (Murray, Smith and Associates). This analysis also benefited greatly from the insights provided by Mark Knudson, who helped formulate the focus of the analysis and provided data for use in quantitatively exploring Portland regionalization issues with WEAP. We greatly appreciate the support of Morteza Anoushiravani, Administrator of the Portland Water Bureau in conducting this study.

ATTACHMENT A.1 – WHOLESALE CONTRACTORS

The Bureau of Water Works has wholesale contracts with the following water purveyors in the Portland metropolitan area.

Contractor	Model node
GNR Corporation	B1
Green Valley Water Company	B1
Gresham, City of	C1
Hideaway Hills Water Company	B1
Lorna Water District	P.V. Lorna
Lusted Water District	B1
Pleasant Home Water District	B1
Powell Valley Road Water District	P.V. Lorna
Rockwood Water People's Utility District	E1
Skyview Acres Water Company	B1
Burlington Water District	Other Demands 2
Palatine Hill Water District	Other Demands 2
Milwaukee, City of	Emergency Only
Lake Grove Water District	Other Demands 3
Tualatin Valley Water District	N1
Beaverton, City of	I1
City of Tigard Water Department	W1 or I1
Valley View Water District	Other Demands 2
West Slope Water District	Other Demands 2
Raleigh Water District	N1
Tualatin, City of	X1

ATTACHMENT A.2 – PROPOSED MINIMUM FLOW REQUIREMENTS

Normal Year

From	To	Minimum (cfs)	% of inflow	Cap (cfs)
1-Jan	31-May	120		
1-Jun	15-Jun	120		
16-Jun	30-Jun	Ramp from 120–35 cfs at start of drawdown		
1-Jul	30-Sep	20–40 cfs, average 35 cfs		
1-Oct	15-Oct	70	50%	400
16-Oct	31-Oct	70	50%	400
1-Nov	15-Nov	150	40%	400
16-Nov	30-Nov	150	40%	400
1-Dec	31-Dec	120		

Critical Year

From	To	Minimum (cfs)	% of inflow	Cap (cfs)
1-Jan	31-May	120		
1-Jun	7-Jun	Ramp from 120–30 cfs if drawdown begins before June 15		
8-Jun	30-Jun	30		
1-Jul	30-Sep	30		
1-Oct	15-Oct	30		
16-Oct	31-Oct	30	50%	250
1-Nov	15-Nov	30	40%	250
16-Nov	30-Nov	70	40%	350
1-Dec	31-Dec	120		

Critical Year Trigger Criteria

Spring: Drawdown before June 15
Summer: Lowest 10% June inflows
Fall: Lowest 10% August and September inflows

APPENDIX B
REGIONAL STRATEGIES FOR MEETING WATER DEMAND IN THE CENTRAL COASTAL PLAIN OF NORTH CAROLINA

Regional Strategies for Meeting Water Demand in the Central Coastal Plain of North Carolina

AwwaRF project 2950:
"Regional Solutions to Developing Water Supplies"

September, 2005

Prepared by
Greg Characklis
University of North Carolina

BACKGROUND

Across the United States, water supplies have become strained by limited supplies and increasing demands for water. Traditionally thought of as a problem in western states, concerns regarding the reliability and sustainability of water supplies in the East have become more pointed in recent years. Developing new water supplies to replace or augment existing supplies can be an expensive process, both monetarily and in terms of the regulatory hurdles that must be cleared. Consequently, utility directors are often reluctant to begin such development until absolutely necessary.

The Central Coastal Plain of North Carolina is a prime example of demand outstripping supply and illustrates many of the challenges and opportunities that are typical of water scarce regions. North Carolina is rife with small water systems, as over 90% of community systems serving populations of fewer than 3,300 people. The Central Coastal Plain (CCP) is a rural region in the eastern third of North Carolina that is dotted with small towns. A group of formations collectively known as the Cretaceous Aquifers supply the vast majority of water to communities in the CCP. The aquifers have a reasonable yield and produce a high quality water that generally requires little, if any, treatment, resulting in an inexpensive water supply.

As population and water demand have increased in the CCP, water levels within the aquifer have been declining. The state has classified drawdown of a foot or more per year to be "significant," a rate that has been exceeded in many regions within the CCP. Declining water levels have resulted in two other related problems. Aquifer dewatering, which occurs when the water level drops below the aquifer's confining layer, can lead to strata compacting and the permanent loss of storage space within the aquifer. Salt water intrusion is also an issue, as a wedge of salt water (water with elevated dissolved solids levels) is moving inward from the coast.

Through out the 1980s and 90s, monitoring data suggested that these problems were growing in severity and that the future viability of the formation was at risk. In response, the state invoked the Water Use Act of 1967, which authorizes the state to declare the affected region a Capacity Use Area (CUA). Once a CUA has been established, the state can promulgate rules that regulate the threatened resource to ensure sustainable usage. The Central Coastal Plain Capacity Use Area (CCPCUA) that was ultimately defined covers an area including all or parts of 15 counties (Figure B.1).

The final CCPCUA rules aim for groundwater withdrawals to be reduced over a three-phase, sixteen-year period (NC DENR 2000). Every groundwater user that consumes more than 10,000 gallons per day (gpd) is required to obtain a permit which entitles it to continue to pump at current, "base level," rates. Larger users, those pumping more that 100,000 gpd will be required to reduce groundwater consumption by anywhere from 10% to 25% (relative to current use) at the end of each of the three phases (6 years, 5 years, 5 years), resulting in total reductions between 30% and 75%. There is no accommodation provided for water systems whose demand rises over the sixteen-year period, so at the end of this time many users will be entitled to withdraw even less than 25% to 70% of their daily demands.

The reasons that make groundwater so attractive within the CCPCUA (low treatment costs, widespread availability) make surface water equally unattractive. There are several rivers within the region that could serve as water sources, but in practical terms, only two, the Neuse and Tar Rivers, are thought of as viable water supplies. In addition, very little surface water treatment capacity in the region, with only one surface water treatment plant on each river (Goldsboro on the Neuse; Greenville on the Tar). Tapping these surface sources will require

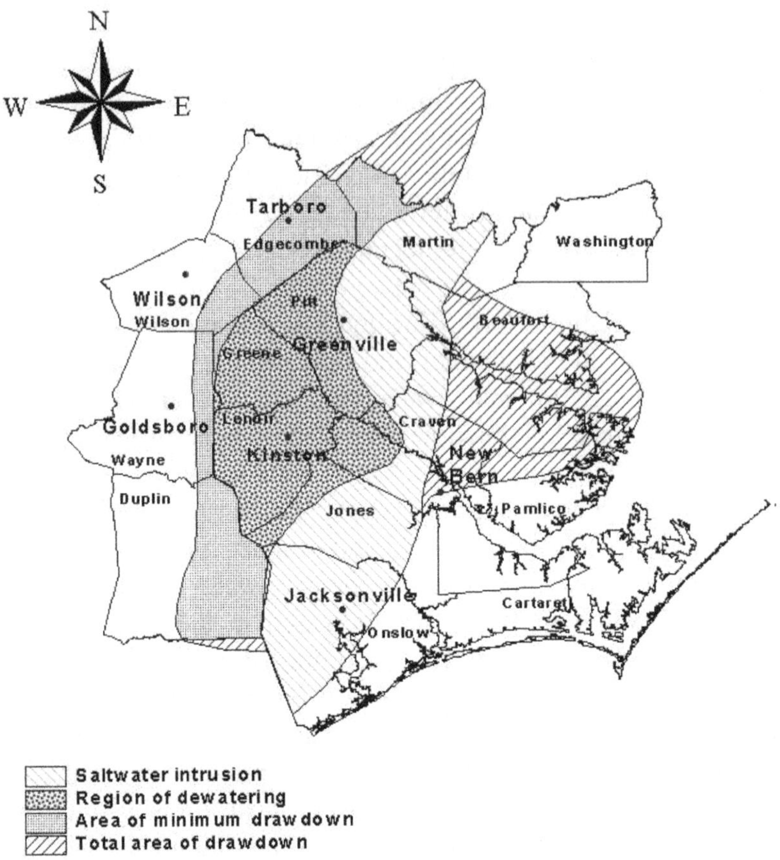

Figure B.1 Map of Central Coastal Plain Capacity Use Area (courtesy North Carolina Department of Environment and Natural Resources)

substantial investments, particularly given that many communities do not lay along the rivers, so conveyance infrastructure will also be required to bring surface water to these groups.

Surface water is not, however, the only alternative available as communities may have access to other supply options. In addition to the Cretaceous Aquifers (Black Creek, Upper Cape Fear, and Lower Cape Fear) there are several other minor aquifers throughout the region that may be available to augment the supplies of some systems. Another alternative, one that has been proposed but not yet tested, is the purchase of groundwater pumping capacity from the Cretaceous Aquifers from those that choose to transition to surface water. The CCPCUA regulations make provisions for the trading of regulated groundwater capacity among permitted users, subject to approval from the North Carolina Department of Environment and Natural Resources (NC DENR).

The North Carolina Department of Environment and Natural Resources, as well as some of the cities within the region, had been monitoring the Cretaceous Aquifers for some time. There concern over the health of the formation was always balanced against the knowledge that whatever steps were taken to protect the aquifers would have a considerable impact on the affected communities. Not only would the communities' primary water supply be severely restricted, but the primary alternative (i.e., surface water) would be quite expensive to develop. The costs of surface water development include substantial investments in both conveyance and treatment infrastructure. In considering alternative strategies for meeting the water demand of these small communities, regionalized facilities began to appear attractive. Surface water treatment exhibits significant economies of scale, with the average cost of produced water ($/thousand gallons, or $/kgal) decreasing rapidly as a plant's capacity increases from that of very small plants (<1 MGD) to around 10 MGD. The vast majority of affected communities in the CCP have demand for less than 2 MGD, as such both NC DENR and local planning assistance agencies (North Carolina Rural Center), encouraged the affected communities to begin considering cooperative approaches to meeting regional potable water demand. Still, a number of challenges remained.

CHALLENGES

First among these according to Nat Wilson, head of the Groundwater Section of the Division of Water Resources at NC DENR, was simply convincing people there is a problem, because, as he said in a recent conversation, "This is a groundwater issue, and groundwater can't be seen." Trying to convince local politicians that they need to spend considerable amounts of money on a problem that they cannot see with their own eyes is a difficult task, according to him. There were, however, areas in which problems had reached a point where they had become self-evident. Kinston is one of the larger cities within the region and the area in and around it has seen some of the largest declines in aquifer levels. "We had enough documentation from 1975 to 1995 to convince us that one day we would turn on the pumps and there would be no water," says Ralph Clark, Kinston's city manager.

The challenge of making the pumping restrictions equitable fell to officials at NC DENR Division of Water Resources, who developed an interesting structure for the cutbacks. One early lesson in developing these plans, however, was the need for transparency. "In our initial attempt at [developing] the rules, we thought we would do a management plan and issue permits, but didn't really communicate how people would be affected. In the second generation of rules, we were more explicit," Wilson says. "Since then, people have been much more knowledgeable about the rules and their situation." Once the communities had a firm understanding of their status, they were able to begin an organized planning process to meet the impending pumping reductions.

In the case of a number of communities, these plans have included the formation of regional surface water systems, which include both regional treatment facilities and conveyance networks (Figure B.2), with the increased scale of the facilities allowing for some reductions in per unit cost. Presently there are two different groups in the CCPCUA taking two separate paths to regionalization. The first group is the Neuse Regional Water and Sewer Authority (NRWASA) which will tap the Neuse River and is centered in Lenoir County. The second group is forming around the city of Greenville which draws water from the Tar River.

With the creation of the CCPCUA, the county commissioners of Lenoir County urged the five water systems in the county to form a single regional system. Through much hard work at every level, plans for NRWASA have progressed. As of July 2005, plans for the treatment plant, to be located a few miles upstream on Kinston on the Neuse, are being prepared to go out to bid. Current expectations are that construction of these facilities will be completed in 2008, the same year in which the first phase of cutbacks are to be enacted. In this case, it made sense to build the system large enough to begin to move these communities entirely off of groundwater, giving each a significant volume of excess permitted pumping capacity. The size of the regional system has

Figure B.2 Map of the affected cities in Central Coastal Plain Capacity Use Area and proposed regional networks (▪ = current/proposed surface water treatment plant)

grown since its initial inception, as the original five Lenoir County members have more recently been joined by four additional water systems from outside the county. The state legislature has also sought to make more alternatives available through legislation allowing private, non-profit water corporations to partner with public water systems in efforts to form regional systems.

One of the biggest proponents of NRWASA has been Ralph Clark, Kinston's city manager. Kinston is easily the largest water system in the NRWASA system and lies directly on the Neuse River. Consequently, Kinston would logically be in the best position to build its own, separate, surface water system, yet Clark sees NRWASA both as a necessity and a win-win situation for all involved. Clark claims that Kinston simply could not have afforded a surface water treatment plant outside of a regional scheme. "How much burden can you put upon 11,000 people?" he asks. As the largest city in NRWASA, Clark feels that Kinston's role is not to dictate terms to the other members, but to help provide the leadership to pull everything together in the end. This is a viewpoint that is easy to understand once it is recognized that every member contributes something. In the case of NRWASA, Kinston owns the permits to the Neuse River water, while several of the smaller systems have access to USDA assistance, programs not available to Kinston.

When asked about the challenges that have arisen as the system has developed, Clark says they are no different than any other situation in which one has to get four or five different groups to agree on something. Addressing this challenge forms the basis of the take home message that he often shares with others. "You have to tell everyone, 'It's a lot of giving and taking.' Every member is equally important." It is a realization that has served NRWASA well, as they have been asked to speak to other communities around the state facing similar groundwater issues.

Greenville has taken a different approach to creating a regional system. Greenville currently treats surface water from the Tar River to meet its demand, and as such, is not directly impacted by the CCPCUA regulations. It is, however, located in the heart of the region, and its 22.5 MGD surface water treatment plant has considerable excess capacity. After the CCPCUA regulations were passed, Greenville invited a number of local utilities to a meeting. The ensuing discussion revolved around Greenville's excess capacity and its willingness to sell wholesale water to those systems that were interested. Since then, two communities, Farmville and Greene County, have signed 40-year contracts in which Greenville agrees to provide them water on an interruptible basis. The agreement has Greenville supplying substantial quantities of water to these communities during periods when it has excess capacity, but it reserves the right to withhold that water up to 36 days per year when it faces peak demands in its own service area. As the groundwater cutbacks are based on a community's average annual consumption, Farmville and Greene County's permitted groundwater capacity following the cutbacks will be adequate to meet their demand when Greenville interrupts the service. At present, all parties seem satisfied with this agreement. Greenville is able to use its excess capacity, while Farmville and Greene County are able to meet their future water demand in spite of the impending withdrawal reductions. Randy Emory, head of Greenville's water utility, cites very few problems in the process. There are some differences among the water systems, such as Greenville's use of chloramines and the groundwater systems' reliance on chlorine as a disinfectant, but plans for pipelines to connect the water systems are progressing. This arrangement has been so successful that other local water systems are now in discussion with Greenville about entering into similar contracts.

Throughout this process, water systems throughout the region have had several resources available to them. In addition to the North Carolina Department of Environment and Natural Resources, the North Carolina Rural Center has worked to provide technical expertise and financial advice to communities facing the prospect of building new water-related infrastructure in the CCP. The Rural Center was established to address the needs of small communities, and has played an active role in assisting the affected communities. Jean Crews-Klein, vice president of business and natural resource development, sees the role of the Rural Center as two-fold: facilitating communication among the parties and providing assistance in the funding process. Crews-Klein notes the challenges facing regional systems by saying, "The most pressing problem is where to get money for a regional system. Getting them started is a huge effort that requires huge capital investment. There is way more need than capital to address [the problem], and it places funders in a difficult position: Do you fund a system that may be having health problems, or do you fund a regional system because it's the right thing to do?" Nonetheless, she expresses pleasure at the amount of progress that has been made in the CCP, progress largely facilitated by thoughtful dialogue among regional cities since the pumping reduction rules were passed. Equally pleased is Don Rayno, community planner in the Division of Water Resources at NC DENR, who says, "We're seeing a fair amount of creativity, which we were hoping to see, and good solutions that weren't anticipated."

RECENT RESEARCH IN REGIONAL SURFACE WATER SYSTEMS

The CCPCUA has spawned research that examines the potential benefits of combining regional surface water systems with tradable groundwater permits as a means of cost-effectively coping with a reduction in groundwater pumping capacity (Kirsch and Characklis n.d.). Regional surface water systems would allow small systems to take advantage of economies of scale, while the sale of their groundwater permits would both offset the costs of joining a regional surface water treatment system and provide a means of augmenting supply for cities located further from surface water sources. In the case of the latter, joining the regional surface water system would be very costly, and it is often much less expensive for these communities to remain entirely on groundwater.

Regionalization is an area that has been examined in the technical literature, with significant research having been done in the area of regional wastewater systems. This body of work primarily involved the use of linear programming models to identify minimum cost regional wastewater configurations (Converse 1972, Deininger and Shaw 1973, Whitlatch and ReVelle 1976, Nakamura et al. 1981, Leighton and Shoemaker 1984). Regional wastewater systems are directly analogous to regional surface water systems, but consideration of a market for groundwater permits dictates a significantly different approach. A deterministic model was created that evaluates both regional surface water systems and tradable groundwater permits as a means of meeting regional demand in the face of groundwater pumping reductions.

The model's premise is that each community is an individual entity that acts in its own best interest and is responsible only for its own costs. The choice of water supply is based entirely upon cost, and the criterion that each city uses to compare water supply alternatives is the unit price increase at the tap ($/kgal). What necessitates a break from previous linear programming research is the fact that a city's cost for joining a regional surface water system is interdependent on the cost of groundwater permits. This results in a circular system in which increased groundwater permit prices lead to more cities joining regional surface water systems, but as more cities join regional surface water systems, demand for groundwater permits drops, driving their cost lower. In order to resolve these circularities, an equilibrium approach was developed to determine each city's least expensive water supply choice given the choices made by every other city in the region.

Evaluating Water Supply Costs

Determining the cost of a water supply alternative requires the determination of numerous components of that cost, particularly for a city joining a regional surface water system. For instance, a conventional surface water treatment plant's cost is estimated according to the following equation:

$$C_{cap}^{plant}(\$) = 2.43 \times 10^{6} \cdot Q_{capacity}^{0.67}$$

where $Q_{capacity}$ = capacity of treatment plant (millions of gallons per day, MGD)

This is a relationship derived from work by Clark and Morand (1981) and updated to 2002 dollars. The capital cost is based on a design capacity corresponding to 1.5 times average demand, which is then amortized and converted into a unit cost ($/kgal). Annual operations and maintenance costs (equation not shown) are based on average annual demand and are also expressed in $/kgal. Conveyance costs are treated similarly, but are assigned to cities within a pipeline network according to the idea that each city is responsible only for its own costs. For instance, in Figure B.3, city B is responsible for all of the capital and operating costs associated with conveyance from city A to city

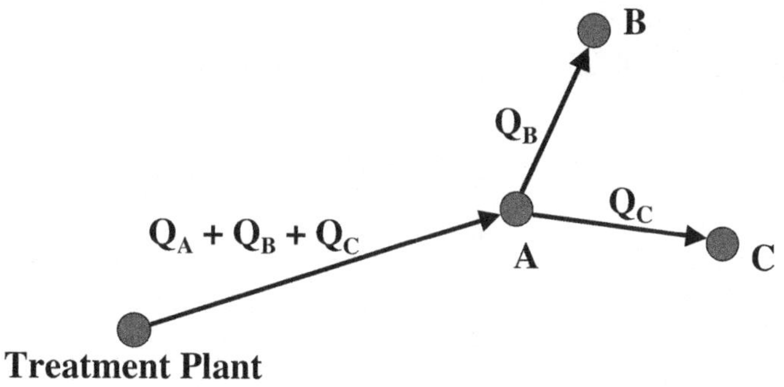

Figure B.3 City B is responsible for all costs from city A to B and for the fraction $Q_B/(Q_A + Q_B + Q_C)$ of costs from the treatment plant to city A

B, as well as those costs corresponding to the fraction of pipe capacity that is devoted to its flows between the treatment plant and city A.

Each city that joins a regional surface water system also obtains revenue from the sale of its groundwater permits, thereby offsetting some of its costs for joining the new system. The total unit cost increase at the tap for a city *i* that is joining a regional surface water system is calculated according to:

$$(Unit\ cost\ increase\ at\ the\ tap)_i = Treatment_{capital} + Treatment_{O\&M} + Conveyance_{capital,I} + Conveyance_{O\&M,I} - GW\ Permit\ Revenues - GW\ Pumping\ Costs$$

Calculating the unit cost of groundwater is more straightforward. No city is having its groundwater capacity eliminated; rather, each city is retaining some fraction, z_i, of its ground-water capacity. Therefore, if the market price of groundwater permits is P_{GW}, the unit cost increase for city *i* following the purchase of sufficient capacity to remain entirely on groundwater would be $P_{GW} \cdot (1 - z_i)$. In this case there is no increase in treatment costs as there is no change from what the utility's current operations.

A conceptual representation of the equilibrium approach used for comparing alternatives is described in Figure B.3 and B.4. The model begins by setting a high groundwater price, such that every city finds it less expensive to join a regional surface water system. As the groundwater price is incrementally lowered, cities will begin to find groundwater less expensive and will drop out of the regional surface water systems. The reduction in capacity due to the one city's dropping out will raise the system-wide costs for other regional system participants, forcing a reconsideration of their earlier choices. The model reaches its endpoint when each city has chosen its least cost alternative and the available groundwater pumping capacity (the permits owned by cities now using surface water) has been fully utilized.

This approach can also include consideration of other sources whose costs are independent of the decisions of others. In the case of the CCPCUA, these are referred to by the term *tertiary alternative*, which includes any supply alternative available only to a single community and whose costs are not related to either the regional systems or groundwater permit market price. Common examples include options such as the drilling of wells into an unregulated surficial aquifer or the construction of a surface water treatment plant serving an individual community.

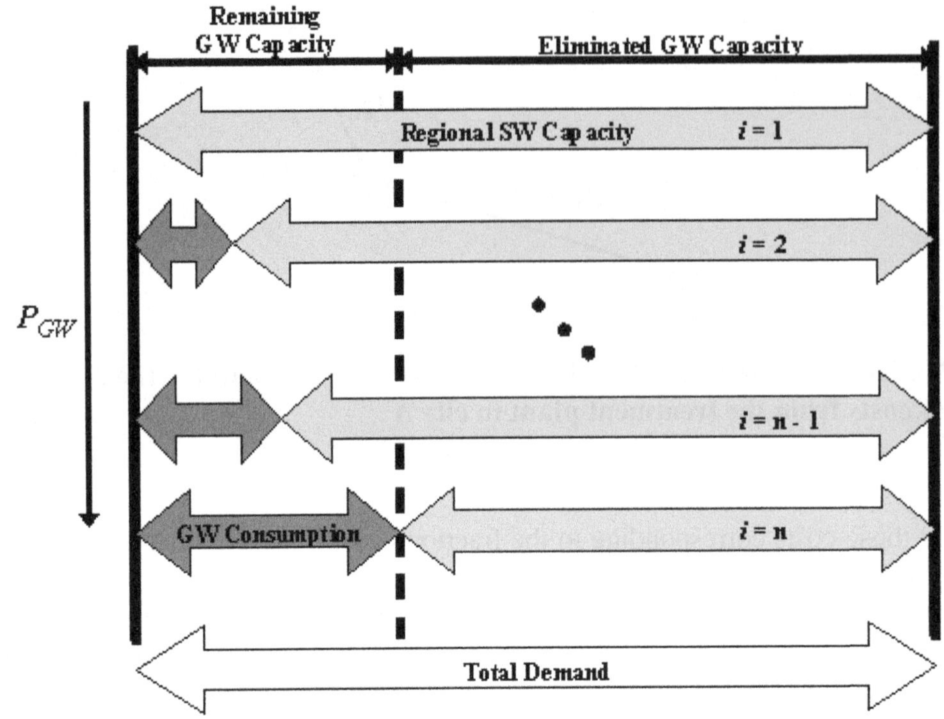

Figure B.4 Conceptual path of the model

This equilibrium approach was designed for application to any region experiencing a reduction in groundwater availability. Data requirements are general, with the required input vectors describing each city's total demand for water, their permitted groundwater consumption, the unit cost of each city's tertiary alternative, and each city's (topographical) elevation. A square matrix describing the distance required for a pipeline from each city to every other city is also required and is used in the development of the pipeline network.

The model allows for multiple surface water systems, each originating from a separate hub to be considered simultaneously within a region. When this approach was applied to the CCP, three regional surface water treatment plant sites were considered: NRWASA, Greenville, and Goldsboro (which operates a surface water treatment plant on the Neuse River). As this is a deterministic model, interruptible contracts were not considered, but expansion of the model to consider stochastic scenarios involving surface water availability have been considered.

Three different scenarios were evaluated, involving regional treatment hubs at Goldsboro, NRWASA and Greenville; NRWASA and Greenville; and Greenville and Goldsboro. Though the model is not a siting model, various combinations of the three regional systems can be considered. Results of those model runs are described in Table B.1, and the individual water supply choices of the affected cities may be seen in Figure B.5.

Based on an independent consultant's report, the estimated capital cost of meeting the CCPCUA regulations without consideration of regional surface water systems or tradable groundwater permits is estimated to be $107 million, with an estimated annual O&M cost of $10 million (Golder and Associates 2002). Assuming a 30-year lifespan and a 5.5% discount rate this results in a regional present value cost of $252.3 million. Model results (Table B.1) suggest that consider-

Table B.1
Results of modeling regional surface water systems, groundwater permit trading, and tertiary alternatives

	Goldsboro—NRWASA—Greenville	NRWASA—Greenville	Greenville—Goldsboro	Limited GW permit trading (3 SW Systems)
Capital cost ($ MM)	$61.4	$66.1	$69.9	$70.4
O&M cost ($ MM / yr)	$7.1	$7.1	$6.6	$8.3
Present value cost* ($ MM)	$164.0	$169.2	$165.4	$191.0
Regional SW users	9	10	9	10
GW purchasers	9	8	9	8
Tertiary users	11	11	11	11
GW purchased (MGD)	4.24	3.79	5.02	1.54

*30 year life with a discount rate of 5.5%.

able cost savings are achieved through the consideration of regional surface water systems and tradable groundwater permits. When Goldsboro, NRWASA, and Greenville are each included, the present value cost is reduced by 35% from the base case. Approximately one-third of the modeled communities are able to meet their groundwater reductions through the purchase of groundwater permits, while in each case at least as many cities join regional surface water systems. Figure B.4 indicates that many of the communities that decide to purchase groundwater permits are those that are in fact closer to the periphery of the CCPCUA and more distant from the regional treatment sites, such that they would face the highest conveyance costs if they were to join a regional surface water system.

The first three columns of Table B.1 assume that groundwater permit trading is unrestricted across the CCPCUA. However, regulators may wish to prevent the net inflow of groundwater permits into areas of particularly heavy aquifer damage. Therefore, the model allows for restrictions to be placed on the sale of groundwater permits across the region. The western third of the CCPCUA has suffered the least aquifer damage and has been demarcated in Figure B.6. Restrictions were placed on the model that limit groundwater permit trading to prevent permits from being traded between areas of high and low aquifer damage. Such restrictions raise the present value cost by $27 million (last column of Table B.1), resulting from the decreased volume of groundwater permits that are allowed to be traded.

As can be seen in both the activities currently ongoing in the CCPCUA and in the model results, regional surface water systems can provide communities with attractive alternatives that can lower the costs of surface water treatment. The economies of scale that regional systems gain are a strong incentive for their creation. In addition, the development of a market for trading groundwater pumping capacity can provide communities with greater flexibility, and even lower costs, as they seek to meet their water supply needs.

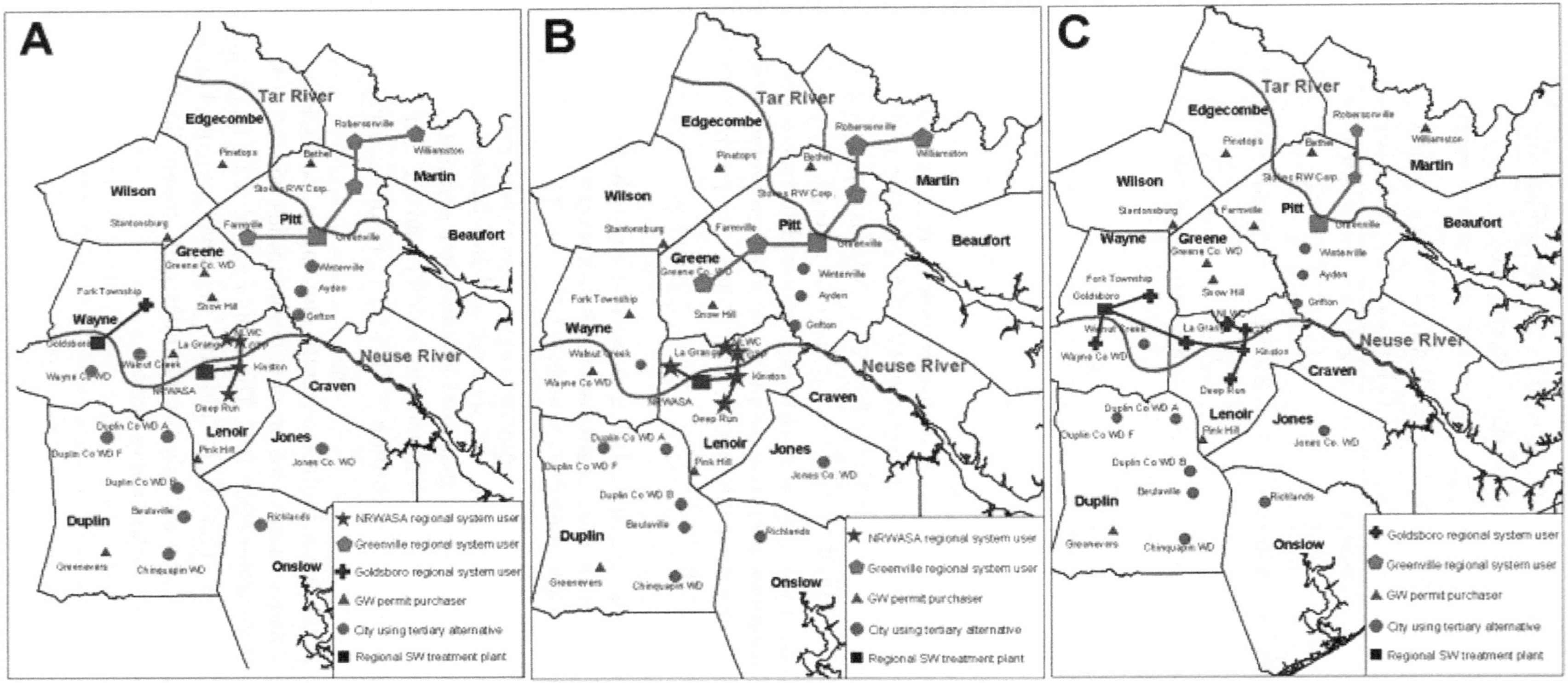

Figure B.5 Results of model runs: (A) considers Goldsboro, NRWASA, and Greenville, (B) includes NRWASA and Greenville, and (C) includes Goldsboro and Greenville

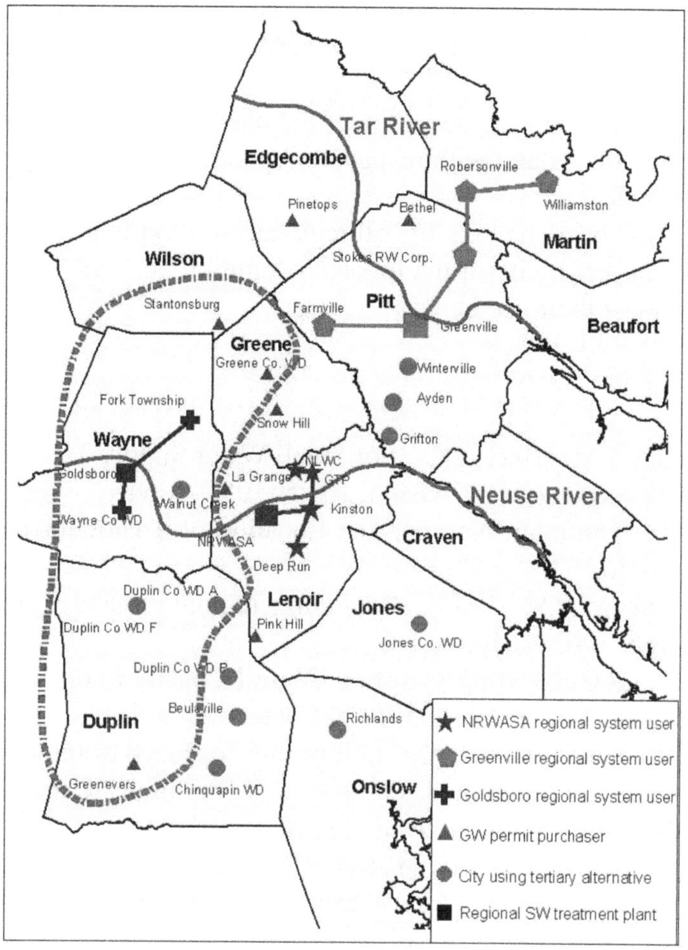

Figure B.6 Restricted groundwater permit trading

LESSONS LEARNED

The process of establishing the CCPCUA and working to meet its requirements has had its difficulties, but the mood of most everyone involved is now one of optimism. Numerous lessons have arisen out of the experiences of the CCPCUA, but as progress continues the overarching lesson from both NRWASA and the Greenville system is that regional approaches can work:

- Regional solutions allow communities to each contribute something and to play to their strengths. Larger cities may have better bond ratings than smaller communities, but the smaller communities may have access to funding sources (e.g., USDA) that are not available to larger systems.
- Taking advantage of economies of scale will be critical in the development of regional systems. While the planning and negotiation associated with developing regional systems will always take time, this approach can result in substantial reductions in unit costs, particularly for small communities.

- The use of markets, such as with the tradable groundwater permits, may provide an efficient means of distributing water rights throughout a region and benefit both buyers and sellers.
- Communication is key at every step of the planning process. Communities must know exactly where they stand with respect to regulators as well as potential partners in any regional scheme.
- There need not be a one-size-fits-all regional solution (e.g., Greenville vs. NRWASA). Ultimately, the creativity of the involved communities will provide them with the best solution to meet their needs.

REFERENCES

Clark, R.M. and Morand, J.M. (1981). "Cost of Small Water Supply Treatment Systems." *Journal of the Sanitary Engineering Division*. 107:5, 1051–1064.

Converse, A.O. (1972). "Optimum Number and Location of Treatment Plants." *Journal WPCF*, 44(8), 1629–1636.

Deininger, R.A., and Shaw, Y.S. (1973). "Modeling regional waste water treatment systems." *Water Research*, 7, 633–646.

Golder and Associates. (2002). "North Carolina Central Coastal Plain Capacity Use Area Water Resources Study." North Carolina Rural Center.

Kirsch, B.R., and G.W. Characklis. n.d. An Equilibrium Approach to Integrating Regional Surface Water Treatment and Limited Groundwater Pumping Capacity. *Journal of Water Resources Planning and Management*. Forthcoming.

Leighton, J.P., and Shoemaker, C.A. (1984). "An Integer Programming Analysis of the Regionalization of Large Wastewater Treatment and Collection Systems." *Water Resources Research*, 20(6), 671–681.

Nakamura, M., and et al. (1981). "Multiperiod design of regional wastewater systems: generating and evaluating alternative plans." *Water Resources Research*, 17(5), 1339–1348.

NC DENR, Division of Water Resources. (2000). "Central Coastal Plain Capacity Use Area Rules."

NC DENR, Division of Water Resources. (2001). "GIS Data of CCPCUA."

State of North Carolina. (1967) "Water Use Act of 1967."

Whitlatch, E.E.J., and ReVelle, C.S. (1976). "Designing Regionalized Waste Water Treatment Systems." *Water Resources Research*, 12(4), 581–591.

REFERENCES

AWWA (American Water Works Association). 2004. CWD History: From Acrimony to Consolidation. *Mainstream.* 1(9) June 8.

AwwaRF (Awwa Research Foundation). 1997. *Guidelines for Implementing an Effective Integrated Resource Planning Process.* Denver, Colo.: AwwaRF and AWWA.

AwwaRF (Awwa Research Foundation). 2003. *Best Practices for Energy Management.* Denver, Colo.: AwwaRF.

Baumann, D., J. Boland, and M. Hanemann. 1997. *Urban Water Demand Management and Planning.* New York: McGraw Hill.

Beecher, J.A., P.C. Mann, and J.D. Stanford. 1993. Meeting Water Utility Revenue Requirements: Financing and Ratemaking Alternatives. Columbus, Ohio: The National Regulatory Research Institute.

Beecher, J.A., and S.J. Rubin. 2000. *Deregulation! Impacts on the Water Industry.* Denver, Colo.: AwwaRF.

Bowers, M.H., and J. Carolan. 1985. The Water Supply System of Metropolitan Boston. Prepared for the Metropolitan District Commission. Wellesley, Mass.: Cultural Resource Group, Louis Berger & Associates, Inc.

Brackman, H., and S.P. Erie. 2001. California's Water Wars Enter a New Century. University of California San Diego [Online]. Available: <http://www.igs.berkeley.edu/publications /par/winter2001/water.htm>. [cited August 4, 2005]

Brookings Institution. 2003. *Back to Prosperity: A Competitive Agenda for Renewing Pennsylvania.* Washington, D.C.: Brookings Institution.

Cadmus Group. 2002. Small Drinking Water Consolidation: Selected State Program and Consolidation Case Studies. Prepared for U.S. Environmental Protection Agency. Boston, Mass.: Cadmus Group Inc.

Castillo, E., S.K. Keefe, R.S. Raucher, and S.J. Rubin. 1997. *Feasibility of Small System Restructuring to Facilitate SDWA Compliance.* Denver, Colo.: AwwaRF and AWWA.

CEC (California Energy Commission). 2005. *First Workshop on the Water-Energy Relationship White Paper.* Integrated Energy Policy Report Workshop. [Online]. Available <http://cacx.org/2005_energypolicy/documents/2005-01-14_workshop/2005-01-14_TRASK_MATT.PDF>. [cited September 15, 2005]

Conservation Law Foundation. 2005. [Online]. Available: <http://www.clf.org>. [cited May 15, 2005]

Cromwell, J.E. III, and S.J. Rubin. 1995. *Development of Benchmark Measures for Viability Assessment.* Harrisburg, Pa.: Pennsylvania Department of Environmental Protection.

Cromwell, J.E. III, and J.L. Jordan. 1999. Linking Full Cost Recovery and Sustainability. In *Providing Safe Drinking Water in Small Systems: Technology, Operations and Economics.* Edited by J.A. Cotruvo, G.F. Craun, and N. Hearne. Washington, D.C.: Lewis Publishers.

Davis, M. 1998. Stepping Outside the Box: Water in Southern California. Speech by Martha Davis at the UCLA Environment Symposium, March 3, 1998 [Online]. Available: <http://www.monolake.org/waterpolicy/outsidebox.htm>. [cited August 2, 2005]

Dedekorkut, A. 2003. Adaptive Governance of Florida's Water Conflicts. Symposium: November 21, 2003. Chapter II: Tampa Bay Water Wars: From Conflict to Collaboration? [Online]. Available: <http://www.fsu.edu/~spap/conference/case/ch2.htm>. [cited August 25, 2005]

Dziegielewski, B., R. Beck, and T. Bik. 2000. Benchmark Investigation of Small Public Water Systems Economics. Research Report. Department of Geography, Southern Illinois University/Carbondale, Illinois [Online]. Available: <http://mtac.sws.uiuc.edu/finalrep.asp>. [cited November 2000]

Eagleton Institute of Politics. 1998. Regionalization: Little Awareness, Little Consensus. *The Star-Ledger.* May 2, 1998 [Online]. Available: <http://slerp.rutgers.edu/retrieve.php?id = 117-6>. [cited March 3, 2004]

Eagleton Institute of Politics. 2004. Regionalization: A Long Row to Hoe. *The Star-Ledger.* January 25, 2004 [Online]. Available: <http://slerp.rutgers.edu/retrieve.php?id = 145-4>. [cited March 3, 2004]

Elkind, S.S. 1998. *Bay Cities and Water Politics: The Battle for Resources in Boston and Oakland.* Lawrence, Kan.: University of Kansas Press.

EMA, Inc. 2004. *Phase I Report prepared for Lehigh County Authority and City of Allentown.* St. Paul, Minn.: EMA, Inc.

Graham, A.G., W.E. Hathhorn, R.L. Wubbena, L.K. Lampe, and N. Grigg. 1999. *Managing Constraints to Water Source Development.* Denver, Colo.: AwwaRF and AWWA.

Grigg, N.S. 1989. Regionalization in Water Supply Industry: Status and Needs. *Journal of Water Resources Planning and Management,* 115(3):367–378.

Herrick, C., J. Pratt, R. Raucher, N. Kalas-Adams, J. Cotruvo, K. Darr-Bornstein, T. Gablehouse, J. Mencer, R. Moser, D. Newkirk, M. Schnack, and K. Westby. 2006. *Emergency Response and Recovery Planning for Water Systems: A Kit of Tools.* Denver, Colo.: AwwaRF.

International Association of Chiefs of Police (IACP). 2003. *Consolidating Police Services: An IACP Planning Approach.* Alexandria, Va.: IACP.

Jesperson, K. 2003. Water System Consolidation Works. *On Tap Magazine* [Online]. Available: <http://www.nesc.wvu.edu/ndwc/articles/OT/SU03/Consolidiation.html>. [cited March 1, 2004]

Kirsch, B.R., and G.W. Characklis. n.d. An Equilibrium Approach to Integrating Regional Surface Water Treatment and Limited Groundwater Pumping Capacity. *Journal of Water Resources Planning and Management. Forthcoming.*

LCA (Lehigh County Authority). 2004. *2003 Annual Report.* Allentown, Pa.: LCA.

LCA (Lehigh County Authority). 2006. *2005 Annual Report.* Allentown, Pa.: LCA.

Macpherson, B. 2005. Three Players Keep the Water Flowing. Tampa Bay Newspapers. August 24 [Online]. Available: <http://www.tbnweekly.com/content_articles/082405_pco-02.txt>. [cited September 14, 2005]

Massachusetts Office of Coastal Zone Management. 2005. 26. Interbasin Transfer Act [Online]. Available: <www.mass.gov/czm/envpermitinterbasinact.htm>. [cited July 6, 2005]

Means, E., and J. Malinowski. 2005. Personal communication, Malcolm Pirnie Inc., August 30, 2005.

Moody's Investors Services. 2000. *Consolidation Will Pressure Ratings of the Water Utilities.* New York: Moody's Investors Services.

MWD (Metropolitan Water District of Southern California). 2005a. Achievements in Conservation, Recycling and Groundwater Recharge [Online]. Available: <http://www.mwdh2o.com/mwdh2o/pages/yourwater/SB60/sb60_main.html>. [cited August 10, 2005]

MWD (Metropolitan Water District of Southern California). 2005b. Aqueduct Magazine—Then and Now [Online]. Available: <http://www.mwdh2o.com/Aqueduct/then_now_1.html>. [cited August 2, 2005]

MWD (Metropolitan Water District of Southern California). 2005c. Bay/Delta [Online]. Available: <http://www.mwdh2o.com/mwdh2o/pages/yourwater/supply/baydelta01.html>. [cited August 4, 2005]

MWD (Metropolitan Water District of Southern California). 2005d. California's Colorado River Allocation [Online]. Available: <http://www.mwdh2o.com/mwdh2o/pages/yourwater/supply/colorado/colorado04.html>. [cited August 4, 2005]

MWD (Metropolitan Water District of Southern California). 2005e. Comprehensive Annual Financial Report, 2004 [Online]. Available: <http://www.mwdh2o.com/mwdh2o/pages/finance/finance01.html>. [cited August 22, 2005]

MWD (Metropolitan Water District of Southern California). 2005f. IID Water Conservation [Online]. Available: <http://www.mwdh2o.com/mwdh2o/pages/yourwater/supply/conservation/conserv02.html>. [cited August 4, 2005]

MWRA (Massachusetts Water Resources Authority). 2002. *MWRA Water System: Supply and Demand* (Draft). Boston, Mass.: MWRA.

MWRA (Massachusetts Water Resources Authority). 2004. Summary Report of MWRA Demand Management Program, Fiscal Year 2003. Boston, Mass.: Massachusetts Water Resources Authority.

MWRA (Massachusetts Water Resources Authority). 2005. News and Reports [Online]. Available: <http://www.mwra.com>. [cited July 10, 2005]

NC DENR (North Carolina Department of Environment and Natural Resources). 2000. *Central Coastal Plain Capacity Use Area Rules. Raleigh, N.C.: NC DENR, Division of Water Resources.*

NRC (National Research Council). 1997. Safe Water from Every Tap: Improving Water Service to Small Communities. Washington, D.C.: National Academy Press.

NRC (National Research Council). 2002. *Privatization of Water Services in the United States: An Assessment of Issues and Experience*. Washington, D.C.: NRC, Water Science and Technology Board.

Okun, D.A. 1977. *Regionalization of Water Management: A Revolution in England and Wales.* London: Applied Science Publishers.

Okun, D.A. 1995. Addressing the Problems of Small Water Systems. In *Proc. of the International Water Supply Association Congress.* Burban, Republic of South Africa. London: International Water Supply Association.

Ottem, T., R. Jones, and R. Raucher. 2003. Consolidation Potential for Small Water Systems—Differences Between Urban and Rural Systems. White paper. Duncan, Okla.: National Rural Water Association.

Peoria Times. 2004. Officials Present Pros and Cons of Consolidation [Online]. Available: <http://peoriatimes.com/articles/2004/02/03/news/news04.txt>. [cited March 3, 2004]

Public Health Service. 1965. Statistical Summary of Municipal Water Facilities in the U.S., January 1, 1963. Public Health Service Publication No. 1039. Washington, D.C.: Public Health Service.

Rand, H. 2003. *Water Wars: A Story of People, Politics, and Power.* Tampa, Fla.: University of South Florida.

Raucher, R.S., D. Chapman, J. Henderson, M.L. Hagenstand, J. Rice, J. Goldstein, A. Huber-Lee, W. DeOreo, P. Mayer, B. Hurd, R. Linsky, E. Means, and M. Renwick. 2005. *The Value of Water: Concepts, Estimates, and Applications for Water Managers.* Denver, Colo.: AwwaRF.

Reese, A.G., R.N. Palmer, and S.E. Nelligan-Doran. 2000. Potential Benefits of Water Supply Regionalization: A Case Study of the Seattle and Everett Water Systems. Presented at ACSE Conference of 2000 [Online]. Available: <http://www.tag.washington.edu/papers/papers/Reese-etal.2000.ASCE-Conf-Proc.0-7844-0517-4.pdf>. [cited June 29, 2004]

Regan, K. 2003. Balancing Public Water Supply Adverse Environmental Impacts Under Florida Water Law: From Water Wars Toward Adaptive Management. *Journal of Land Use.* Vol. 19:1 [Online]. Available: <www.law.fsu.edu/journals/landuse/vol19_1/regan.pdf>. [cited August 12, 2005]

Rubin, S.J. 2001. Criteria to Assess the Affordability of Water Service. White paper. Duncan, Okla.: National Rural Water Association.

Rubin, S.J. 2002. Overview of Small Water System Consolidation. Presented to National Drinking Water Advisory Council (NDWAC) Small Systems Affordability Working Group, Washington, D.C., October 21, 2002.

Shanahan, P. 2002. Trends in Cooperation, Barriers To It, and Strategies for Overcoming Barriers and Encouraging Cooperation. Presented to NDWAC Small Systems Affordability Work Group, Washington, D.C., October 21, 2002.

SMC Martin. 1983. Regionalization Options for Small Water Systems. Prepared for USEPA, Office of Drinking Water (Contract 68-01-6285), Valley Forge, Pa.

State of Georgia. 2004. Consolidating Services for Efficiency [Online]. Available: <http://www.dca.state.ga.us/publications/csindex.html>. [cited March 3, 2004]

Stratus Consulting, Inc. 2000. *Economic Outlook Study for Areawide Water and Wastewater Service.* Prepared for the Lehigh County Authority. Boulder, Colo.: Stratus Consulting, Inc.

TBW (Tampa Bay Water). 2001. *Tampa Bay Water's Master Water Plan.* Tampa, Fla.: TBW.

TBW (Tampa Bay Water). 2005. Welcome to Tampa Bay Water [Online]. Available: <http://www.tampabaywater.org>. [cited August 31, 2005]

USEPA (U.S. Environmental Protection Agency). 1993. Technical and Economic Capacity of States and Public Water Systems to Implement Drinking Water Regulations. EPA810-R-93-001. Washington, D.C.: USEPA.

USEPA (U.S. Environmental Protection Agency). 1994. Joining Forces on Solid Waste Management: Regionalization is Working in Rural and Small Communities. EPA530-K-93-001. Washington, D.C.: USEPA.

USEPA (U.S. Environmental Protection Agency). 1995. Unpublished data from the Safe Drinking Water Information System. Washington, D.C.: USEPA.

USEPA (U.S. Environmental Protection Agency). 1999. National Characteristics of Drinking Water Systems Serving Populations Under 10,000. EPA816-R-99-010. Washington, D.C.: USEPA.

USEPA (U.S. Environmental Protection Agency). 2002a. FY 2002 Public Water System Inventory Data. Data from SDWIS/FED [Online]. Available: <http://www.epa.gov/safewater/data/detaileddetails>. [cited March 3, 2003]

USEPA (U.S. Environmental Protection Agency). 2002b. System Partnership Solutions to Improve Public Health Protection [Online]. Available: <http://www.epa.gov/safewater/smallsys/pdfs/capacitydevelopstudyv15.pdf>. [cited March 1, 2004]

USEPA (U.S. Environmental Protection Agency). 2003. 40 CFR Parts 141, 142, and 143. National Primary Drinking Water Regulations: Stage 2 Disinfectants and Disinfection Byproducts; Proposed Rule. *Fed. Reg.* (Part II) 68(159): 49547–49681.

USEPA (U.S. Environmental Protection Agency). 2006. National Primary Drinking Water Regulations: Stage 2 Disinfectants and Disinfection Byproducts; Final Rule. *Fed. Reg.* (Part II) 71(2): 387–493.

USGS (U.S. Geological Survey). 1995. Washington Water Science Center [Online]. Available: <http://wa.water.usgs.gov?wuse/main.cntv.95.text>. [cited July 8, 2005]

Vaux, H.J. Jr. 1988. Growth and Water in the South Coast Basin of California. In *Water and Arid Lands of the Western United States*. Edited by M.T. El-Ashry and D.C. Gibbons. New York: Cambridge University Press.

Wade Miller Associates. 1991. State Initiatives to Address Non-viable Small Water Systems in Pennsylvania. Harrisburg, Pa.: Pennsylvania Department of Environmental Resources, Bureau of Community Environmental Control, Division of Water Supplies.

Wakstein, A.M. 1972. Boston's Search for a Metropolitan Solution. *AIP Journal*, 38(September): 285–296.

Wallace, Floyd Associates. 1984. A History of the Development of the Metropolitan District Commission Water Supply: Water Supply Study and Environmental Impact Report. Prepared for Metropolitan District Commission. Boston, Mass.: Wallace, Floyd Associates.

ABBREVIATIONS

AWWA	American Water Works Association
AwwaRF	Awwa Research Foundation
CCP	Central Coastal Plain
CCPCUA	Central Coastal Plain Capacity Use Area
CIWA	Central Iowa Water Authority
CoP	City of Portland
CUA	Capacity Use Area
CWA	Clean Water Act
CWD	Champlain Water District
CWS	community water system
DBP	disinfection by-product
DOH	Department of Health
DWSRF	Drinking Water State Revolving Fund
ESA	Endangered Species Act
FRDS	Federal Reporting Data System
GAC	granular activated carbon
GCMS	gas chromatography mass spectrometry
GIS	geographic information systems
GMA	Growth Management Act
gpd	gallons per day
HUD	Housing and Urban Development
IID	Imperial Irrigation District
IRP	integrated research planning
LCA	Lehigh County Authority
MCL	maximum contaminant limit
MDC	Metropolitan District Commission
mgd	million gallons per day
MWD	Metropolitan Water District of Southern California
MWRA	Massachusetts Water Resources Authority
MWRC	Massachusetts Water Resources Commission
NC DENR	North Carolina Department of Environment and Natural Resources
NEWWA	New England Water Works Association
NRC	National Research Council

NRWA	National Rural Water Association
NRWASA	Neuse Regional Water and Sewer Authority
NT-NCWS	non-transient non-community water systems
O&M	operation and maintenance
OSHA	Occupational Safety and Health Administration
PAWC	PA American Water
PSC	Public Service Commission
PUC	Public Utility Commission
PWB	Portland Water Bureau
RUS	Rural Utility Service
SDWA	Safe Drinking Water Act
SDWIS	Safe Drinking Water Information System
SRF	state revolving fund
SWFWMD	Southwest Florida Water Management District
SWOT	Strengths Weaknesses Opportunities Threats
TBW	Tampa Bay Water
TMF	technical, management, and financial
U.S.	United States
USDA	U.S. Department of Agriculture
USEPA	U.S. Environmental Protection Agency
UV	ultraviolet
WAC	wastewater advisory panel
WCRWSA	West Coast Regional Water Supply Authority
WEAP	water evaluation and planning systems
WERF	Water Environment Research Foundation
WSCAC	Water Supply Citizens Advisory Committee

IWA Publishing's authorised EU representative for General Product Safety Regulations is Diane D'Arras, 15 rue Duret, 75116 Paris, France, e-mail: safety@iwap.co.uk.

Printed and bound by CPI Group (UK) Ltd, Croydon, CR0 4YY
18/09/2026
02218938-0002